ENCYCLOPAEDIA
OF
MINERALOGY

ENCYCLOPAEDIA
OF
MINERALOGY

Volume 2

Bharat Singh

ANMOL PUBLICATIONS PVT. LTD.
NEW DELHI-110 002 (INDIA)

ANMOL PUBLICATIONS PVT. LTD.
Regd. Office: 4360/4, Ansari Road, Daryaganj,
New Delhi-110002 (India)
Tel.: 23278000, 23261597, 23286875, 23255577
Fax: 91-11-23280289
Email: anmolpub@gmail.com
Visit us at: www.anmolpublications.com

Branch Office: No. 1015, Ist Main Road, BSK IIIrd Stage
IIIrd Phase, IIIrd Block, Bengaluru-560 085 (India)
Tel.: 080-41723429 • Fax: 080-26723604
Email: anmolpublicationsbangalore@gmail.com

Encyclopaedia of Mineralogy

ISBN 978-81-261-3475-5 (Set)

PRINTED IN INDIA

Printed at AnVi Composers, New Delhi.

Contents

Preface

'Mineralogy' is a branch of science, under which we study about the minerals, which are obtained from the earth. All the minerals are found in the earth. We dig out the earth for getting minerals in the form of ores, which are impure. By the process of Metallurgy, we get pure elements. A few metals such as Au (gold), Ag (silver) and Pt (platinum) are found in pure form. A mineral is defined to be a naturally occurring, homogeneous crystalline solid. First, a mineral should be a solid; this requirement excludes all liquids and gases. A crystalline material is not merely a solid, but also possesses an orderly and periodic molecular structure. In recent years, however, a multitude of synthetic materials, which would otherwise satisfy the definition of a mineral, have been produced in the laboratory. These substances are rapidly gaining in importance in both industry and experiment, and it now seems logical to extend the definition of 'mineral'. Minerals are inorganic solids, which have a regular three-dimensional ordered structure of atoms.

Minerals are classified on the basis of their chemical composition and their isomorphism or similarity of properties, such as structural properties, physical properties and chemical properties.

In structural properties, we study about the structure of the crystal. Crystallographic property is one of the most characteristic properties of the crystals, and in this property we study about the shape of the crystals. Other structural property is 'optical properly', in which we study about the crystal structure, the kind of atoms present in the structure and their electronic arrangement. In physical properties, we study about the properties such as colour, taste, melting and boiling points, density, etc. In chemical properties, we study about the reactions of the elements/minerals with others and products after reactions and their relation with the reactants.

This eminent work, *Encyclopaedia of Mineralogy* has specific and realistic features about minerals in an interesting manner, which will be able to make a creative impression in the study of mineralogy. Expectedly, this work would be welcomed and liked by all.

— **Editor**

1

Classification of Mineral Species

Berzelian Classification

The Berzelian mineral classification system was named in honour of the Swedish chemist and mineralogist Jons Jakob Berzelius (1779-1848). The Berzelian system categorises mineral species according to the main anion group present in their chemical structure. All mineral species of a certain class are therefore chemically similar because they possess the same main anion group. Mineral classes may then be further subdivided according to physical features, which cations are present, the presence or absence of water or the hydroxyl anion, or internal structure.

The main classes which are recognised under Berzelius' scheme include the native elements; sulphides and sulphosalts; oxides and hydroxides; halides; carbonates, nitrates and borates; sulphates; phosphates; and silicates. The antimonides, arsenides, selenides, and tellurides closely resemble the sulphides in composition, while the chromates, molybdates and tungstates resemble the sulphates. The arsenates and vanadates are closely akin to the phosphates.

The native elements include all mineral species which are composed entirely of atoms in an uncombined state. Such minerals either contain the atoms of only one element or else are metal alloys. The native elements are divided into metallic, semi-metallic, and non-metallic subgroups. Metals tend to be dense and malleable substances which possess a characteristic metallic lustre and conduct electricity well. Semi-metals and non-metals are brittle and conduct poorly compared to metals.

Minerals of the sulphide class are compounds which contain the non-metallic element sulphur in combination with atoms of a metal or a semi-metal. Compounds in which anions of antimony (Sb), arsenic (As), selenium (Se), or tellurium (Te) replace the sulphur anion and bond with metallic or semi-metallic cations are classed respectively as antimonides, arsenides, selenides, and tellurides. If the sulphur anion, a metallic element, and a semi-metal are all present then the mineral is categorised as one of the rare sulphosalts. Most sulphides and sulphosalts are soft, dark, heavy, and brittle, possessing a distinct metallic lustre and high conductivity.

The minerals of the oxide class are those which contain oxygen bonded to one or more metallic elements. Hydroxides are compounds of a metallic element and water or the hydroxyl anion $(OH)^-$. The oxide minerals tend to be relatively hard, and some of them may be used as gemstones. Many provide economically important metal ores. Minerals of the hydroxide class tend to be softer and less dense than oxides.

In members of the halide class an element of the halogen group such as fluorine (F), chlorine (Cl), bromine (Br), or iodine (I) bonds to a metal or semi-metal cation such as sodium (Na), potassium (K), magnesium (Mg), calcium (Ca), aluminium (Al), copper (Cu), or silver (Ag). Halides are constructed entirely of ionic bonds. The halide minerals tend to be soft, brittle, easily soluble in water, and possess medium to high melting points. They are poor conductors when in solid state.

Mineral species which are members of the carbonate class are compounds of a metal or semi-metal with the carbonate anion $(CO_3)^{2-}$. The plane triangular carbonate anion groups thus form the basic unit from which carbonate minerals are constructed. The nitrates are structurally very closely akin to the carbonates. Nitrogen bonds to three oxygen atoms to form the nitrate radical, $(NO_3)^-$, which forms the basic building block of the minerals of this species. The nitrates tend to be softer and possess lower melting points than the carbonates. Atoms of the element boron (B) join to three oxygen atoms in order to form the borate radical, $(BO_3)^{3-}$. This anion group closely resembles the carbonate and nitrate radicals in structure. However, the oxygen atoms of the borate radical may, unlike those of the carbonate or nitrate radicals, be shared between anion groups. Borate radicals may therefore be linked into polymerised chains, sheets, or multiple groups. These are the chemical structures which compose the minerals of the borate class.

The sulphur cation may form very strong bonds with four oxygen atoms, producing the anion group $(SO_4)^{2-}$. This sulphate radical forms the basic structural unit of the minerals of the sulphate class. The sulphate radical does not share oxygen atoms and cannot polymerise. Minerals of the chromate class are compounds of metallic cations with the chromate anion group $(CrO_4)^{2-}$. Just as sulphur and chromium form the anion groups $(SO_4)^{2-}$ and $(CrO_4)^{2-}$, the ions of molybdenum (Mo) and tungsten (W) bond with oxygen atoms to create the anion groups $(MoO_4)^{2-}$ and $(WO_4)^{2-}$. These anion groups then bond with metal cations to form the minerals of the molybdate and tungstate classes. Molybdenum and tungsten may freely substitute for one another within the ionic groups $(MoO_4)^{2-}$ and $(WO_4)^{2-}$, allowing the formation of series of solid solution. They may not, however, substitute for sulphur within the sulphate radical $(SO_4)^{2-}$ or form solid solution with minerals of the sulphate or chromate classes. Species of the molybdate and tungstate classes are typically heavy, soft, and brittle. They tend to be dark or vividly coloured.

Like sulphur, the elements phosphorous (P), arsenic (As), and vanadium (V) form tetrahedral anion groups in combination with oxygen. The resulting phosphate radical, $(PO_4)^{3-}$, provides the basic structural unit of the minerals of the phosphate class; the arsenate and vanadate radicals $(AsO_4)^{3-}$ and $(VaO_4)^{3-}$ form the basic structural units of the arsenate and vanadate classes. The mineral species of these three classes are thus composed of the respective phosphate, arsenate, and vanadate radicals linked by various metal and semi-metal cations. Phosphate, arsenic and vanadium ions may substitute for one another within the three anion groups, forming series of solid solution.

The basic constituent of the minerals of the silicate class is the silicate radical $(SiO_4)^{4-}$. Each oxygen atom within a silicate radical may bond with another silicon ion, becoming part of a second silicate radical and linking the two radicals together. One, two, three, or four of the oxygen atoms in each silicate anion group may bind to other silicate tetrahedra in this way. Many different structures are therefore possible; silicate radicals may remain structurally isolate, join together in pairs, or link into frameworks, sheets, chains, or rings. The various species of the silicate class are grouped according to their structural type. Silicate minerals are usually of relatively great hardness, and single crystals are often translucent.

Native Elements

The native elements are those minerals formed wholly from elements which occur in an uncombined state. No ionic or covalent bonding may join atoms of one element to atoms of another within the lattice structure of such a mineral. Usually only one type of atom is present in the molecular structure of these species. However, the metal alloys, which contain two or more metals in solid solution, are also classified as native elements because the different species of atom present in their lattices are joined only by metallic bonds, not by ionic or covalent bonds. The native elements are further categorised into subgroups containing metals, semi-metals, and non-metals.

METALS

Metallic elements which are found in the native state include gold, silver, copper, lead, iron, nickel, platinum, and the rarer elements-palladium, iridium, and osmium. Mercury, tantalum, tin, and zinc have also been found. The uncombined atoms of the metals act as perfect spheres and are relatively inert; they tend to form lattices of face-centred cubic, body-centred cubic or hexagonal close-packed structure. The lattice structures of these native metals are composed of metallic bonds, which are relatively weak and produce soft, malleable, ductile, and sectile substances with rather low melting points. Because many electrons are free to move about within the lattice the native metals are very conductive, and because light cannot propagate inside a good conductor the metals possess the characteristic highly reflective 'metallic' lustre.

The most common native metals are members of the gold group. These include the elements gold (Au), silver (Ag), copper (Cu), and lead (Pb). Mineral species composed of uncombined atoms of the gold group elements possess face-centred cubic lattices, which schema is also known as cubic closest packing. This cubic closest packing results in a high number of atoms per unit volume and thus in a mineral of relatively high density. In such lattices there exist no atomic planes distinguished by higher or lower density; minerals of the gold group therefore demonstrate hackly fracture and possess no cleavage.

Gold and silver atoms possess equal 1.44 angstrom atomic radii. These two substances are thus mutually soluble, and occur in mixtures with a wide range of relative compositions. Copper, however, has a radius of 1.28 angstrom. For this reason it is present only in tiny amounts within mixtures of gold and silver and conversely contains only tiny amounts of gold and silver in solution.

The iron group contains iron (Fe) and nickel (Ni). These metals possess a body-centred cubic structure. Nickel and iron are mutually soluble because their atomic radii are both equal to 1.24 angstrom. This solid solution is frequently found in meteorites and probably constitutes much of the earth's core.

Platinum (Pt), palladium (Pd), iridium (Ir), and osmium (Os) number among the less common metallic native elements and are classified as the platinum group. The mineral species which form from the uncombined atoms of these elements are platinum, palladium, platiniridium and iridosmine. Platiniridium is a rare alloy of iridium and platinum, while iridosmine is an equally rare alloy of iridium and osmium. Both alloys possess hexagonal close-packed structure, while platinum and palladium are cubic close packed. The metals of the platinum group are harder and possess higher melting points than those of the gold group.

Semi-metals

The native semi-metals include arsenic (As), antimony (Sb), and bismuth (Bi), as well as the less common elements selenium (Se) and tellurium (Te). Arsenic, antimony and bismuth crystallise in the hexagonal-scalenohedral class while selenium and tellurium crystallise in the trigonal-trapezoidal class. Natural crystals of all three species are rare. The semi-metals are brittle, and conduct heat and electricity poorly compared to the metals. However, like the metals they display a metallic lustre.

The lattices of the semi-metals are composed of bonds intermediate in type between metallic and covalent. Such bonds are stronger than metallic bonds but are also more directional. The structure of the hexagonal semi-metals is therefore based on a distorted form of cubic closest packing in which sheets of atoms parallel to the base of the crystal separate into pairs.

This sheetlike structure results in perfect basal cleavage, or perfect cleavage along the paired planes.

The distorted form of cubic closest packing found in the semi-metal lattice possesses a greater volume than the close packing which composes the liquid form of the substances. The semi-metals thus display the unusual property of expansion upon crystallisation.

Non-metals

The native non-metals include carbon (C), in the form of diamond and graphite, and sulphur (S).

Polymorphism in Diamonds and Graphites

The mineral species diamond and graphite offer a spectacular example of the trait of polymorphism. Polymorphism occurs when two or more mineral species contain exactly the same elements in exactly the same proportions, and therefore share a chemical formula, yet possess dissimilar lattice structures. Two polymorphic minerals possess identical chemical formulae but different crystal structures; the minerals may therefore exhibit very different physical traits.

Chemically the two species diamond and graphite are identical. Both are native elements composed wholly of elemental carbon; both may be burned to carbon dioxide at high temperatures; and both contain closely similar carbon-carbon bonds. However, the two substances are structurally very different. Diamond possesses an exceptionally strong lattice in which each carbon atom is bonded by four covalent bonds to four neighbouring carbon atoms, which occupy the apices of a regular tetrahedron. All four valence electrons are taken up by covalent bonds, so that none are free to conduct electricity; hence diamond forms a highly successful insulator. Graphite, on the other hand, is composed of sheets of six-atom rings in which each carbon has three neighbouring atoms positioned at the corners of an equilateral triangle. Three valence electrons are occupied by covalent bonds; the fourth is free to act as a conductor of electricity. Graphite thus conducts relatively well. The sheets are stacked a distance much greater than one angstrom apart, and the van der Waals forces which bind the stacked sheets together are very weak. The wide separation and weak binding forces between parallel sheets result in perfect basal cleavage.

The two non-metallic native elemental species graphite and diamond possess identical chemical formulae but, because of their differing lattice structures, demonstrate widely disseparate physical traits. They thus provide an apt example of polymorphism.

Sulphides, Antimonides, Arsenides, Selenides, Tellurides, and Sulphosalts

Sulphides: Sulphide minerals are compounds of one or more metal or semi-metal elements with the non-metallic element sulphur (S). In a sulphide, the sulphur anion (S^{2+}) is thus combined with metallic cations such as iron (Fe), cobalt (Co), nickel (Ni), copper (Cu), zinc (Zn), molybdenum (Mo), silver (Ag), cadmium (Cd), tin (Sn), platinum (Pt), gold (Au), mercury (Hg), tellurium (Tl), lead (Pb), and the semi-metals arsenic (As), antimony (Sb), and bismuth (Bi). Compounds of sulphur and one of the semi-metals are termed semi-metal sulphides. Most sulphides are structured by ionic bonds, although some may contain metallic bonding and display metallic properties.

ANTIMONIDES, ARSENIDES, SELENIDES, TELLURIDES

The non-metal sulphur (S) and the semi-metals selenium (Se)·and tellurium (Te) consistently play the electronegative, anionic role in their compounds. However, the semi-metals arsenic (As) and antimony (Sb) may either play the anionic role often filled by sulphur or else act as a cation by donating electrons and forming a compound with sulphur.

When the semi-metals antimony (Sb), arsenic (As), selenium (Se), and tellurium (Te) occupy the role of a non-metal and substitute for the sulphur anion, the minerals formed possess a chemical structure very similar to that of the sulphides. These are classified respectively as antimonides, arsenides, selenides, and tellurides according to which semi-metal is present. When the semi-metals arsenic (As) and antimony (Sb) form compounds with sulphur, however, they occupy the role of a metal cation and the minerals which contain them are termed semi-metal sulphides. If a metal, a semi-metal, and non-metallic sulphur are all present then the mineral is categorised as one of the rare sulphosalts.

If both a metal and a semi-metal such as arsenic or antimony are present in a compound and there is no sulphur anion present, then the semi-metal replaces sulphur in order to play the electronegative role. In this case the mineral is classified as an antimonide, arsenide, selenide, or telluride. For example, niccolite (NiAs) contains the metal nickel and the semi-metal arsenic, which acts as an non-metal and provides the anion. Niccolite is therefore an arsenide. The mineral breithauptite (NiSb) contains nickel and the semi-metal antimony, which acts as the electronegative, non-metallic element of the compound. Breithauptite is thus an antimonide. The mineral calaverite ($AuTe_2$), in turn, contains the metallic element gold and the semi-metal tellurium, which provides an anion; calaverite is a telluride.

If only sulphur and either arsenic or antimony are present then the semi-metal acts as a cation by donating electrons as it combines with the sulphur. For example, the mineral realgar (AsS) is a sulphide which contains only the semi-metal arsenic (As) and the non-metal sulphur (S). In this case, the arsenic acts as a metal by donating electrons to the electronegative sulphur. This type of compound is considered a semi-metal sulphide; other examples are orpiment (As_2S_3) and bismuthinite (Bi_2S_3).

Sulphosalts

The rare minerals which are compounds of sulphur (S), a semi-metal such as arsenic (As) or antimony (Sb), and one or more metals are termed sulphosalts. In these species the semi-metal plays the role of a metal. The mineral cobaltite, which possesses the chemical formula CoAsS, is a sulphosalt. The metallic element in this compound is cobalt (Co); the semi-metal, which also donates electrons, is arsenic (As), and the non-metal sulphur receives the electrons donated by both the cobalt and the arsenic. Pyrargyrite ("dark ruby silver", chemical formula Ag_3SbS_3), is also a sulphosalt. Its metallic element is silver (Ag), the semi-metal is antimony (Sb), and the non-metal is sulphur. Proustite (Ag_3AsS_3; "light ruby silver") is another sulphosalt with a chemical formula and structure very similar to that of pyrargyrite, differing only because the semi-metal antimony (Sb) has been replaced by arsenic (As).

Most of the sulphides contain ionic bonding. Others contain a degree of metallic bonding. Typical sulphides and sulphosalts are soft, dark, heavy, and brittle, possessing a distinct metallic lustre and high conductivity. Most are opaque, demonstrating the distinctive colours and coloured streak derived from the presence of chromophores (iron (Fe), cobalt (Co), nickel (Ni), and copper (Cu)). Those sulphides which are not opaque (cinnabar, realgar, and orpiment) are transparent only in very thin sections. Some species such as pyrite emit a sulphurous odour when they are struck with a mallet.

Categorisation by Ratio of Metals to Non-metals

The sulphides, antimonides, arsenides, selenides, tellurides, and sulphosalts may be grouped according to the ratio of metal to non-metal contained in their chemical formulae. Let the letter A represent a metal or a semi-metal acting as a cation, while X represents a non-metal or a semi-metal acting as an anion. Then the formulae may be categorised as AX, AX_2, A_2X, and so on according to the ratio of metal to non-metal contained. Galena (PbS), sphalerite (ZnS), chalcopyrite ($CuFeS_2$), covellite (CuS), and cinnabar (HgS) are examples of sulphides whose formulae possess the ratio AX. In all of these the non-metallic element sulphur bonds to various metals such as lead, zinc, copper, iron, and mercury. In the semi-metal sulphide realgar (AsS) arsenic fills the role of a metal and bonds with sulphur. Niccolite (NiAs) offers an example of an AX-type arsenide, in which arsenic plays the non-metal, electronegative role while nickel donates electrons. Breithauptite (NiSb) in turn is an AX antimonide in which antimony fills the non-metal role and bonds with nickel. Chalcocite (Cu_2S) provides an example of an A_2X sulphide, while pyrite (FeS_2), and molybdenite (MoS_2) are sulphides possessing formulae of the AX_2 type. The mineral species cobaltite (CoAsS) is a sulphosalt and calaverite ($AuTe_2$) is a telluride of the AX_2 type. Species of the A_2X_3 type include the semi-metal sulphides orpiment (As_2S_3) and bismuthinite (Bi_2S_3) while the sulphosalts pyrargyrite (Ag_3SbS_3) and proustite (Ag_3AsS_3) are of the form A_4X_3.

AX		
galena	PbS	sulphide
sphalerite	ZnS	sulphide
chalcopyrite	$CuFeS_2$	sulphide
covellite	CuS	sulphide
cinnabar	HgS	sulphide
realgar	AsS	semi-metal sulphide
niccolite	NiAs	arsenide
breithauptite	NiSb	antimonide
A_2X		
chalcocite	Cu_2S	sulphide
AX_2		
pyrite	FeS_2	sulphide
cobaltite	CoAsS	sulphosalt
molybdenite	MoS_2	sulphide
calaverite	$AuTe_2$	telluride
AX_3		
skutterudite	$CoAs_3$	sulphide
A_2X_3		
orpiment	As_2S_3	semi-metal sulphide
bismuthinite	Bi_2S_3	semi-metal sulphide
A_4S_3		
cobaltite	CoAsS	sulphosalt
pyrargyrite	Ag_3SbS_3	sulphosalt
proustite	Ag_3AsS_3	sulphosalt

Secondary Enrichment of Sulphides

Many of the sulphide minerals are water soluble. Such minerals can be dissolved and transported underground by heated groundwater. This hydrothermal transportation of minerals is the mechanism by which secondary enrichment occurs. Sulphides and arsenides are frequently altered to produce the secondary sulphates and arsenates.

Oxides and Hydroxides

Simple and Multiple Oxides: Members of the oxide class are minerals in which an oxygen anion is combined with one or more metals. The oxides can be divided into two categories consisting of simple oxides, which contain a single metallic element, and multiple oxides, which are compounds of oxygen and two or more metals.

Simple oxides are compounds of a single metallic element and oxygen. Examples of such species include zincite (ZnO), tenorite (CuO), cuprite (Cu_2O), rutile (TiO_2), uraninite (UO_2), corundum (Al_2O_3), haematite (Fe_2O_3), and magnetite (Fe_3O_4). Ice (H_2O) provides an unusual example of a simple oxide in which hydrogen replaces a metal in order to act as the cation.

Only one type of metal is present in combination with oxygen in each simple oxide species. Multiple oxides, in contrast, are compounds of oxygen with two or more metallic elements. Examples are ilmenite ($FeTiO_3$), spinel ($MgAl_2O_4$), chromite ($FeCr_2O_4$), and chrysoberyl ($BeAl_2O_4$). Many of these species are colourful, relatively hard, and may be used as gemstones.

The oxide class contains several metal ores of great economic importance. Among these are the iron ores, haematite and magnetite; chromite, an ore of chromium; manganite ($MnO(OH)$), which provides manganese; zincite, which contains zinc; and gibbsite ($Al(OH)_3$), which offers a source of aluminium.

The oxides can be usefully divided into the two categories of simple oxides, which contain only a single cation, and multiple oxides, which contain two or more metals. However, the oxides may alternatively be subdivided so that the chemical formula of each mineral species is categorised according to the ratio of metal to oxygen which it contains. If the capital letter A represents a metal atom, then the formulae AO, A_2O, AO_2, A_2O_3, and A_3O_4 each represent the chemical formulae of a group of mineral species. Examples of the AO type are provided by zincite (ZnO) and tenorite (CuO), while cuprite (Cu_2O) offers an example of the A_2O type. Species of the AO_2 type are rutile (TiO_2) and uraninite (UO_2). The A_2O_3 type is represented by corundum (Al_2O_3), haematite (Fe_2O_3), and ilmenite ($FeTiO_3$), while the A_3O_4 type is represented by spinel ($MgAl_2O_4$), magnetite (Fe_3O_4), chromite ($FeCr_2O_4$), and chrysoberyl ($BeAl_2O_4$).

AO — zincite (ZnO), tenorite (CuO)

A_2O — cuprite (Cu_2O), ice (H_2O)

AO_2 — rutile (TiO_2), uraninite (UO_2)

A_2O_3— corundum (Al_2O_3), haematite (Fe_2O_3), ilmenite ($FeTiO_3$)

A_3O_4— spinel ($MgAl_2O_4$), magnetite (Fe_3O_4), chromite ($FeCr_2O_4$), chrysoberyl ($BeAl_2O_4$)

Hydroxides

Hydroxides are compounds of metallic elements with water or the hydroxyl anion $(OH)^-$. Examples of hydroxides are manganite (MnO(OH)), goethite (FeO(OH)), and gibbsite ($Al(OH)_3$; one of the main components of bauxite). Minerals of the hydroxide class are typically softer than oxides and are of low to medium density.

In terms of chemical composition, quartz (SiO_2) is a member of the oxide class. However in terms of molecular structure, which in the Berzelian classification system is considered to be more fundamental than chemical composition, quartz is classified as a silicate.

Halides

The halide class is composed of minerals in which an element of the halogen group such as fluorine (F), chlorine (Cl), bromine (Br), or iodine (I) provides the anion. These elements form relatively large, lightly charged, easily polarised ions. The halogen ions then bond to large, weakly polarised cations of sodium (Na), potassium (K), magnesium (Mg) or calcium (Ca), and sometimes aluminium (Al), copper (Cu), or silver (Ag). Halides are therefore constructed entirely of ionic bonds. Both cations and anions behave as almost spherical bodies which pack in a highly symmetric manner; most common halites possess isometric hexoctahedral crystalline structure.

The halides tend to be soft, brittle, easily soluble in water, and possess medium to high melting points. They are poor conductors of heat and electricity when in solid state but good conductors when molten. (When molten, ions are liberated and may then transport charges. Molten halides thus conduct by electrolysis, or the transportation of charge by free-moving ions.)

Example— members of the halide class are the mineral species halite (NaCl; also known as rock salt), sylvite (KCl), fluorite (CaF_2), and chlorargyrite (AgCl, also known as horn silver).

Carbonates

Mineral species which are members of the carbonate class are compounds of a metal or semi-metal with the carbonate anion $(CO_3)^{2-}$. In these substances plane triangular $(CO_3)^{2-}$ anion groups are linked together by various cations. Each oxygen atom is bonded more strongly to its associated carbon than to any other atom of the structure, and oxygen atoms are not shared between the carbonate anions. The plane triangular carbonate anions thus form the basic unit from which carbonate minerals are constructed.

The bond between the carbon and the two oxygen atoms

of the $(CO_3)^{2-}$ anion is strong. However, when brought in contact with the hydrogen ion (H^+) the carbonate radical decomposes, producing carbon dioxide and water. Minerals of the carbonate class thus react easily with acids such as hydrochloric acid (HCl). For example, calcite (calcium carbonate, $CaCO_3$) effervesces when placed in an aqueous solution of HCl, producing carbon dioxide and calcium chloride:

$$CaCO_{3\,(s)} + 2HCl_{(aq)} \longrightarrow CaCl_{2\,(s)} + CO_{2\,(g)} + H_2O_{(l)}$$

This reaction provides a means for the identification of carbonate species which is easily applicable in the field.

Anhydrous Carbonates

The anhydrous carbonates lack both the hydroxyl anion (OH^-) and water (H_2O) in their chemical formulae. These carbonates are divided into three isostructural groups, called the calcite group, the dolomite group, and the aragonite group.

The calcite group contains calcite ($CaCO_3$), magnesite ($MgCO_3$), rhodocrosite ($MnCO_3$), and siderite ($FeCO_3$). Each of these minerals is composed of the carbonate anion and cations of a single metal. Although the bonds within the carbonate anion are covalent, the bonds between the carbonate anion and the metal cation are ionic. The structure of the minerals of the calcite group is composed of alternate layers of metal cations and carbonate anions. All members of the group are isostructural, possessing unit cells which are completely analogous in structure and differ only in the identity of the constituent atoms. Because the calcite minerals are isostructural, they form solid solutions of varying proportions in which magnesium (radius 0.66 angstrom), manganese (0.80 angstrom) and iron (0.74 angstrom) substitute for one another. Calcium, however, does not substitute so readily for these elements because of its larger radius (0.99 angstrom). Members of the calcite group are also called the rhombohedral carbonates, because the unit cell and the cleavage both form rhombohedra.

The dolomite group contains the minerals dolomite

($CaMg(CO_3)_2$) and ankerite ($CaFe(CO_3)_2$). Each of these species contains the semi-metal calcium and a metal cation together with the carbonate anion. The structure of the dolomite minerals is composed of layers in which carbonate anions alternate with metal cations. In this way members of the dolomite group are similar in structure to the calcite group. However, the dolomites are more highly ordered than the calcites because their cation layers consist of layers of calcium alternating with layers of magnesium or iron.

Due to its large atomic radius (0.99 angstrom) in comparison to the radii of iron (0.74 angstrom) or magnesium (0.66 angstrom) calcium is unable to form a solid solution with these metals, and the calcium separates out from the iron and magnesium to produce its own layer. Like the members of the calcite group, those of the dolomite group adhere to the rhombohedral crystal system.

The aragonite group is composed of the orthorhombic carbonates, of which aragonite ($CaCO_3$), strontianite ($SrCO_3$) and cerussite ($PbCO_3$) provide examples. The members of this group are isostructural and may form a solid solution in which strontium, lead, and calcium may substitute for one another.

Like the members of the calcite group, the aragonite minerals each contain the carbonate anion and ions of a single metal. However, the members of the aragonite group adhere to the orthorhombic crystal system rather than the rhombohedral crystal system preferred by the calcite and dolomite groups. Structurally calcite and aragonite are thus quite different.

Polymorphism and Pseudomorphism in Calcite and Aragonite

The two different mineral species calcite and aragonite both possess the chemical formula $CaCO_3$. Although chemically identical, these two species are structurally very different; calcite crystalises according to the rhombohedral crystal system while aragonite crystallises according to the orthorhombic crystal system. Aragonite is harder and has a higher specific gravity than calcite. Two mineral species such as calcite and aragonite which are chemically identical and yet differ structurally and possess different physical properties are said to be polymorphs.

The structure of a carbonate is determined by the radius of its constituent cation. Those carbonates which contain smaller cations form rhombohedral structures while those possessing larger cations occupy orthorhombic structures. Magnesium, manganese, and iron are small cations, and the carbonates such as magnesite, rhodocrosite, and siderite which contain them adhere to the rhombohedral crystal system. Lead and strontium, on the other hand, are large cations, and species such as strontianite and cerussite which contain them belong to the orthorhombic crystal system.

The radius of calcium occupies a unique position near the critical radius which distinguishes between the two possible crystal systems. Carbonate species which contain the calcium anion may thus crystallise according to either structural system. For this reason the compound calcium carbonate ($CaCO_3$) crystallises with rhombohedral crystal structure as calcite and with orthorhombic structure as aragonite. The fact that the calcium cation is of approximately the critical radius leads to the ability of calcium carbonate to crystallise in either system. This trait explains the polymorphism of calcite and aragonite, both of which possess the chemical formula $CaCO_3$ but which crystallise according to the differing rhombohedral and orthorhombic crystal systems.

Aragonite is less stable than calcite and therefore less common. Aragonite begins to invert to calcite at temperatures as low as 400° Celcius; in contact with water of aqueous solutions containing calcium carbonate it may even invert at room temperatures. For this reason pseudomorphs of calcite after aragonite are often observed. In such cases the unit cells of the original aragonite have been gradually replaced by those of calcite while the original crystal shape remains intact.

Hydrous Carbonates

The hydrous carbonates consist of the copper carbonates, which contain both copper and the hydroxyl radical (OH^-). Also contained within this subgroup are the hydrated carbonates such as thermonatrite (Na_2CO_3. H_2O) and Trona ($Na_3H(CO_3)_2$. $2H_2O$, which contain water (H_2O) within their lattice structures.

The copper carbonates azurite ($Cu_3(CO_3)_2(OH)_2$), malachite ($Cu_2CO_3(OH)_2$), and aurichalcite ($Cu_5(CO_3)_2(OH)_6$) contain both copper and the hydroxyl anion (OH^-). The element copper is a strong pigmenting agent and renders azurite a bright, distinctive azure blue, malachite a vivid green, and aurichalcite green to blue. Both azurite and malachite are of the monoclinic crystal system while aurichalcite is of the orthorhombic crystal system. Malachite frequently forms a pseudomorph of azurite; azurite is less stable than malachite, and gradually inverts to it over long periods of time.

Nitrates

The nitrates are chemically closely akin to the carbonates. Nitrogen bonds to three oxygen atoms in a structure very similar to the carbonate anion group $(CO_3)^{2-}$. This nitrate radical, $(NO_3)^-$, forms the basic building block of the nitrates.

Because nitrogen is more electronegative than carbon, the nitrate anion group $(NO_3)^-$ is less stable than $(CO_3)^{2-}$. Mineral species of the nitrate class thus tend to occur less frequently in geologic formations than do the carbonates. $(NO_3)^-$ is less charged than $(CO_3)^{2-}$; the nitrates tend therefore to be softer than the carbonates and also to possess lower melting points. The nitrates yield oxygen as the nitrate anion decomposes upon heating.

Several mineral species of the nitrate class form structures analogous to certain of the carbonates. For example, nitratine (or soda nitre; $NaNO_3$) and the carbonate calcite ($CaCO_3$) are isostructural and possess the same rhombohedral cleavage.

(Nitratine is softer than calcite and melts on a lower temperature, both of which are expected of a nitrate as compared to a carbonate). Nitre (KNO_3, known colloquially as saltpeter) is in turn isostructural with the orthorhombic carbonate aragonite ($CaCO_3$). To complete the analogy with calcite and aragonite, nitratine ($NaNO_3$) possesses a polymorph which is also of chemical formula $NaNO_3$. The polymorph of rhombohedral nitratine is isostructural with orthorhombic nitre (KNO_3). Nitratine's polymorphism is thus analogous to that of the rhombohedral carbonate calcite and the orthorhombic carbonate aragonite, both of chemical formula $CaCO_3$.

Borates

The boron atom joins to three oxygen atoms to form the borate radical, $(BO_3)^{3-}$. After this anion group has formed, each oxygen atom still has one electron available for bonding. Thus the oxygen atoms of the borate radical may, unlike those of the carbonate or nitrate radicals, be shared between anion groups. Borate radicals may therefore be linked into polymerised chains, sheets, or multiple groups. The most common borate minerals are composed of sheets of radicals which are linked together through the sharing of all three oxygen atoms. The sheets are then separated by layers of water molecules (H_2O) and linked together by sodium (Na^+) or calcium (Ca^{2+}) ions.

Two example members of the borate class are borax, which possesses the chemical formula $Na_2B_4O_7.10H_2O$, and colemanite, $Ca_2B_5O_{11}.5H_2O$.

Sulphates

The sulphur anion (S^{2-}) may bond to a positive metallic or semi-metallic ion, receiving two electrons in order to fill its valence shell. Mineral species of the sulphide class are compounds of this nature. Alternately, the six electrons of sulphur's unfilled valence shell may be ejected, resulting in the small, highly charged (S^{6+}) cation. The sulphur cation may then form very strong bonds with four oxygen atoms, producing the anion group $(SO_4)^{2-}$. This sulphate radical forms the basic structural unit of the minerals of the sulphate class.

All members of the sulphate class contain the tetrahedral sulphate radical $(SO_4)^{2-}$. These tightly bound anion groups do not share oxygen atoms and cannot polymerise.

Anhydrous Sulphates

Anhydrous sulphates contain neither water molecules (H_2O) nor the hydroxyl anion (OH^-) within their lattice structures. The anhydrous sulphates consist of the barite group and the mineral species anhydrite ($CaSO_4$). The members of the barite group include barite ($BaSO_4$), celestite ($SrSO_4$), and anglesite ($PbSO_4$). These species are isostructural and may form a solid solution by exchanging atoms of strontium and lead for those of barium. The minerals of the barite group all possess orthorhombic symmetry, as does anhydrite. Anhydrite is not, however, isostructural with the barite minerals because the calcium ion possesses a radius smaller than that of barium, strontium, or lead.

Barite ($BaSO_4$) sometimes produces groups of divergent tabular crystals known as "desert roses." The barite mineral anglesite ($PbSO_4$) is formed through the oxidation of lead-bearing veins and in particular the alteration of the lead sulphide galena (PbS). Specimens of concretionary habit have been found consisting of concentric layers often containing a core of unaltered galena.

Hydrous Sulphates

Gypsum ($CaSO_4.2H_2O$) is the most prevalent and important of the hydrous sulphates. It possesses a sheetlike structure which consists of layers of calcium ions and sulphate anions separated by water molecules (H_2O). Gypsum demonstrates perfect cleavage along planes parallel to its sheetlike layers. If the water is driven out of the gypsum structure, the mineral will collapse into the configuration of anhydrite ($CaSO_4$) with a marked decrease in volume and loss of the perfect cleavage. Gypsum is frequently formed during the alteration of anhydrite.

Chalcanthite ($CuSO_4.5H_2O$, also known as blue vitriol), melanterite ($FeSO_4.7H_2O$, known colloquially as copperas), epsomite ($MgSO_4.5H_2O$), antlerite ($Cu_3(OH)_4SO_4$), linarite ($PbCu(SO_4)(OH)_2$), cyanotrichite ($Cu_4Al_2SO_4(OH)_{12}$. $2H_2O$), and alunite ($KAl_3(OH)_6(SO_4)_2$; known as potassium alum) also belong to the hydrous sulphates.

Chalcanthite, antlerite, linarite, and cyanotrichite are copper ores which are found in the oxidised and hydrothermally altered regions of copper veins. Due to the pigmenting effect of copper, chalcanthite, linarite and cyanotrichite are a deep azure blue in colour. Antlerite may be emerald to very dark green and has a pale green streak. All four minerals may form as the product of secondary alteration of copper sulphides such as chalcocite (Cu_2S). Melanterite is an iron ore which is formed by the oxidation and hydrothermal alteration of pyrite (FeS_2) or other iron sulphide minerals. This species is yellow green to deep green due to the presence of iron.

Chromates

Minerals of the chromate class are compounds of metallic cations with the chromate anion group $(CrO_4)^{2-}$. The species of this class of minerals are very rare.

The lead chromate crocoite ($PbCrO_4$) is isostructural with the sulphates of the barite group ($BaSO_4$). This rare mineral is found in the oxidised zones of lead-bearing veins, particularly in places where lead veins have encountered rocks containing chromite. Due to the strong pigmenting capabilities of chromium, crocoite is a bright orange-red in colour while its streak is orange.

Molybdates and Tungstates

Just as sulphur and chromium form the anion groups $(SO_4)^{2-}$ and $(CrO_4)^{2-}$, the ions of molybdenum (Mo) and tungsten

(W) bond with oxygen atoms to create the anion groups $(MoO_4)^{2-}$ and $(WO_4)^{2-}$. These anion groups then bond with metal cations to form the minerals of the molybdate and tungstate classes. However, the atomic radii of molybdenum and tungsten are larger than that of sulphur. For this reason the ionic groups containing molybdenum and tungsten do not form regular tetrahedra. Instead, the tetrahedra occupied by the metal cation and the oxygen atoms are flattened and deformed. Molybdenum and tungsten possess equal atomic radii and may therefore freely substitute for one another within the ionic groups $(MoO_4)^{2-}$ and $(WO_4)^{2-}$, allowing the formation of series of solid solution. Molybdenum and tungsten may not, however, substitute for sulphur within the sulphate radical $(SO_4)^{2-}$ or form solid solution with minerals of the sulphate class.

Minerals containing the molybdate or tungstate anion groups are categorised as members of the molybdate or tungstate classes. These two classes are then further subdivided into the wolframite and scheelite groups. Minerals of either class may be placed into each group according to the relative size of the metallic cation which they contain. Species of the molybdate and tungstate classes are typically heavy, soft, and brittle. They tend to be dark or vividly coloured.

Mineral species belonging to the wolframite group contain small cations such as magnesium, manganese, iron, cobalt, and nickel. The tungstates huebnerite ($MnWO_4$) and ferberite ($FeWO_4$) are considered to be end members of a series of solid solution in which manganese may substitute for various quantities of iron and vice versa. Both huebnerite and ferberite belong to the wolframite group.

Mineral species belonging to the scheelite group form compounds containing larger ions such as calcium. The tungstates scheelite ($CaWO_4$) and stolzite ($PbWO_4$) and the molybdates powellite ($CaMoO_4$) and wulfenite ($PbMoO_4$) are all members of the scheelite group. The lead molybdate wulfenite is a bright red colour and is found in the oxidised portion of lead veins.

Because molybdenum and tungsten possess equal atomic radii, they may freely substitute for one another within the ionic groups $(MoO_4)^{2-}$ and $(WO_4)^{2-}$. One series of solid solution exists between powellite ($CaMoO_4$) and scheelite ($CaWO_4$), while another occurs between wulfenite ($PbMoO_4$) and stolzite ($PbWO_4$). However, despite the similarity in atomic radii the difference in atomic weights between the two elements is significant. Molybdenum possesses an atomic mass of 96 amu, while tungsten's mass is 184 amu. For this reason the two elements are often separated out by the force of gravity during the formation of the primary minerals which bear them. In primary mineral species, therefore, molybdate and tungstate species usually occur separately rather than in mixture. The two elements are more often found in solid solution in secondary minerals which result from the alteration of primary minerals.

Phosphates, Arsenates, and Vanadates

The apatite group contains mineral species of each of the phosphate, arsenate, and vanadate classes. Among the minerals of the apatite group number the three lead ores pyromorphite ($Pb_5Cl(PO_4)_3$), mimetite ($Pb_5Cl(AsO_4)_3$), and vanadinite ($Pb_5Cl(VO_4)_3$). Each of these species contains both lead and chlorine. The three minerals are isostructural, differing only according to whether they contain phosphor, arsenic, or vanadium within the radical. Pyromorphite is thus a phosphate, mimetite is an arsenate, and vanadinite is a vanadate. These three species form a complete series of solid solution in which arsenic and vanadium may replace phosphor. The mineral vanadinite is bright orange-red due to the presence of the chromophore vanadium; all three species are found in the oxidised areas of lead veins.

The apatite group also contains the three species fluorapatite ($Ca_5F(PO_4)_3$), chlorapatite ($Ca_5Cl(PO_4)_3$), and hydroxylapatite ($Ca_5(OH)(PO_4)_3$). These species are all calcium phosphates, because they contain the calcium cation (Ca^{2+}) and the phosphate radical (PO_4^{3-}). However, each species contains a different anion. Fluorapatite contains fluorine (F^-), chlorapatite contains chlorine (Cl^-), and hydroxylapatite contains the hydroxyl anion (OH^-). The three minerals are isostructural, differing only according to which anion fills the requisite space in the unit cell. They form a complete series of solid solution in which chlorine or the hydroxyl anion may replace fluorine.

Examples of phosphates which do not belong to the apatite group class are, amblygonite ($LiAlF(PO_4)$), the hydrous phosphate wavellite ($Al_3(PO_4)_2(OH)_3$), and the hydrous iron phosphate vivianite ($Fe_3(PO_4)_2.8H_2O$). Vivianite is colourless before exposure to light; after exposure to light it assumes a blue or green colour due to the presence of iron. Its streak is blue. Tirquoise ($CuAl_6(PO_4)_4(OH)_8.2H_2O$) is a hydrous copper and aluminium phosphate which may be sky blue, bright blue or tirquoise blue in colour and is pigmented by copper. Lazulite ($(Fe,Mg)Al_2(OH)_2(PO_4)_2$) is a ferric and magnesium phosphate which is a characteristic azure blue in colour.

The mineral species erythrite ($Co_3(AsO_4)_2.8H_2O$ is a hydrous cobalt arsenate. Erythrite is a vivid crimson or pink in colour due to presence of the pigmenting agent cobalt; its streak is pale red or pink. The mineral is often found in the form of characteristic pink crusts called "cobalt bloom". Erythrite occurs as a secondary product of cobalt arsenides such as skutterudite ($CoAs_3$). Nickel may substitute for the cobalt in erythrite to form the mineral annabergite ($Ni_3(AsO_4)_2.8H_2O$), sometimes called "nickel bloom". Annabergite is isostructural with erythrite; both minerals are hydrous arsenates and differ only according to whether they contain cobalt or nickel. Annabergite is a secondary mineral derived from nickel arsenides such as niccolite (NiAs). Nickel, like cobalt, is a chromophore: due to its presence, annabergite is light green in colour.

Conichalcite ($CaCu(AsO_4)(OH)$) is a copper arsenate which is grass green in colour and possesses a green streak. It is pigmented by copper. Olivenite ($Cu_2AsO_4(OH)$) is another copper arsenate which is usually olive green.

The descloizite series includes the vanadates descloizite ($PbZnVO_4(OH)$) and mottramite ($PbCu(VO_4(OH))$). Both minerals are lead vanadates, containing the vanadium radical $(VaO_4)^{3-}$ and a lead cation; they differ in that descloizite contains a zinc cation whereas mottramite contains a copper cation. Descloizite and mottramite are isostructural, forming a series of solid solution in which mottramite's copper substitutes for descloizite's zinc. Mottramite ($PbCu(VO_4)(OH)$) is grass green or olive green in colour due to the pigmenting action of copper; it possesses a green streak. Descloizite and mottramite are found in the oxidised portions of lead, copper, and vanadium veins.

SILICATES

The fundamental constituent of the minerals of the silicate class is the silicate radical $(SiO_4)^{4-}$. This anion group is composed of one silicon atom bonded to four oxygen atoms, which occupy the apices of a regular tetrahedron. The bond which holds the silicon and oxygen atoms together within the silicate radical is partially ionic and partially covalent. This bond originates in the attraction of oppositely charged ions for one another, but it also involves the sharing of electrons and significant overlap of electron clouds.

The various species of the silicate class are grouped according to their structural type. The silicate anions may exist as isolated tetrahedra, paired tetrahera, single or double chains, rings, sheets, or three-dimensional boxlike structures. Groups of minerals containing such structures are given the respective terms nesosilicates, sorosilicates, inosilicates, cyclosilicates, phyllosilicates and tectosilicates.

Minerals of the silicate class are responsible for providing the bulk material out of which the earth's crust and mantle are formed. (Silicates form 95 per cent of the crust and 97 per cent of the mantle). Silicate minerals are usually of relatively great hardness, and single crystals are often translucent.

Polymerisation

The bond between silicon and oxygen utilises only one of oxygen's available bonding electrons. It is therefore possible that each oxygen atom within a silicate radical may bond with another silicon ion, becoming part of a second silicate radical and cementing the two disseparate ion groups together.

One, two, three, or four of the oxygen atoms in each silicate radical may bind to different external silicon anion groups in this way. (Two or more oxygen atoms are never shared between the same adjacent silicate tetrahedra.) This bonding system opens up many structural possibilities; silicate radicals may remain structurally isolate, join together in pairs, or link into rings, chains, sheets, or frameworks. The process of linking series of anion groups together into chains, sheets, or ring structures through the sharing of oxygens is called polymerisation.

In general, the higher the temperature at which a mineral specimen formed the lower the amount of polymerisation which will be present. The silicate minerals in igneous rock undergo a definite sequence of crystallisation beginning with those rocks possessing the least amount of polymerisation and ending with those which possess most.

The element aluminium plays a prominent role in the construction of silicate minerals. Aluminium may, like silicon, join to four oxygen atoms; however, it may also bond with six oxygens. An aluminium atom joined to four oxygen ions (AlO_4) occupies the same volume of space which is needed by a silicate $(SiO_4)^{4-}$ radical; the aluminium anion group may therefore replace the silicate anion within chains or polymers. When joined to six oxygen atoms, however, aluminium can instead cement separate silicate tetrahedra together by forming ionic bonds. The ionic bonds thus formed are much weaker than the ionic-covalent bonds within the silicate and aluminate radicals. Aluminium is not the only element which may bond with oxygen and help to join silicate radicals together. The elements magnesium (Mg), iron (Fe), manganese (Mn), aluminium (Al), and titanium (Ti) may all occur joined to six oxygen atoms within the structure of the silicate minerals.

NESOSILICATES

The nesosilicates or island silicates contain isolated $(SiO_4)^{4-}$ tetrahedra. In the minerals of this group there is no direct linkage between separate silicate anion groups; the silicate tetrahedra are held together only by ionic bonds with various metallic cations.

Examples of the nesosilicates are provided by members of the olivine group, which contains forsterite ($Mg_2(SiO_4)$) and fayalite ($Fe_2(SiO_4)$); as well as members of the garnet group; zircon ($Zr(SiO_4)$); and topaz ($Al_2(SiO_4)(F,OH)_2$).

The garnets provide an interesting example of an isostructural group. The minerals of this group possess the general chemical formula $A_3B_2(SiO_4)_3$. The A lattice site is occupied by large ions such as calcium (Ca), magnesium (Mg), iron (Fe), and manganese (Mn) which possess two bonding electrons. The B lattice site is dedicated to smaller ions possessing three bonding electrons such as aluminium (Al), iron (Fe), titanium (Ti), and chromium (Cr). The magnesium, iron, and manganese inhabitants of the A lattice site are completely interchangeable because they possess almost equal atomic radii. Calcium, however, possesses a differing atomic radius and does not substitute for these three elements. Substitution between aluminium, iron and chromium occurs within the B lattice site.

The pyralspite garnets all have aluminium in the B lattice position; magnesium, iron and manganese may substitute for one another in the A lattice site to produce pyrope ($Mg_3Al_2(SiO_4)_3$), almandite ($Fe_3Al_2(SiO_4)_3$), and spessartite ($Mn_3Al_2(SiO_4)_3$). The ugrandite garnets all contain calcium in the A lattice position; aluminium, iron, and chromium may substitute for one another in the B lattice position. The ugrandite subgroup thus contains grossularite ($Ca_3Al_2(SiO_4)_3$), andradite ($Ca_3Fe_2(SiO_4)_3$), and uvarovite ($Ca_3Cr_2(SiO_4)_3$). The six garnet species are isostructural, possessing lattice configurations which are completely analogous and differ only in the identity of their constituent atoms.

Sorosilicates

In the sorosilicates or couplet silicates two silicate tetrahedra are linked into a pair by sharing a single oxygen ion. The most prevalent sorosilicates are members of the epidote group and are characterised by the mineral epidote, $Ca_2(Al,Fe)Al_2O(SiO_4)(Si_2O_7)(OH)$.

Cyclosilicates

The cyclosilicates or ring silicates are composed of $(SiO_4)^{4-}$ tetrahedra linked into closed cyclic rings by the sharing of oxygen atoms. These rings possess a silicon to oxygen ratio of 1:3. Three possible ring configurations exist; these are the Si_3O_9 ring, observed only in the rare titanosilicate benitoite ($BaTiSi_3O_9$); the Si_4O_{12} ring; and the Si_6O_{18} ring, which is found in beryl and tourmaline. The Si_6O_{18} rings are hexagonal and arranged in planar sheets. In beryl these parallel sheets are so strongly bonded together by beryllium and aluminium cations that they demonstrate only poor cleavage.

Two example members of the cyclosilicate group are beryl ($Be_3Al_2(Si_6O_{18})$), of which varieties are known as aquamarine, morganite, and emerald, and tourmaline ($Na(Mg,Fe)_3Al_6(BO_3)_3(Si_6O_{18})(OH,F)_4$).

Inosilicates

The silicate anion groups of the inosilicates or chain silicates are linked together into single or double chains. In order to form single chains each silicate radical must share two out of its four oxygen atoms with neighbouring radicals. In order to form double chains, half of the silicate radicals share three oxygens rather than two, while the other half continue to share two; the extra bonds affix two chains together into a double chain. Sheaves of parallel single or double chains of silicate tetrahedra are then bound together by metallic cations such as calcium (Ca), magnesium (Mg), or iron (Fe).

Those inosilicate species which are composed of single chains are classified as the pyroxenes, while those which are constructed of double chains are termed the amphiboles. Both groups may contain the same set of cations, which means that many pyroxenes are analogous to a species of the amphibole group and vice versa. For example, the pyroxene hypersthene ($(Mg,Fe)_2(Si_2O_6)$), which contains magnesium and iron in solid solution, is analogous to the amphibole anthophyllite ($(Mg,Fe)_7(Si_8O_{22})(OH)_2$), which contains the same cations. The difference between the two species lies in the ratio of silicon to oxygen, which is a legacy of the respective single or double chain structures, and in the presence of the hydroxyl radical (OH^-) characteristic of amphiboles.

Examples of pyroxene species include, the minerals enstatite ($Mg_2(Si_2O_6)$), diopside ($CaMg(Si_2O_6)$), spodumene ($LiAl(Si_2O_6)$), and jadeite ($NaAl(Si_2O_6)$). The amphiboles include tremolite ($Ca_2Mg_5(Si_8O_{22})(OH)_2$), and glaucophane ($Na_2Mg_3Al_2(Si_8O_{22})(OH)_2$). Wollastonite ($Ca(SiO_3)$) and rhodonite ($Mn(SiO_3)$) are also inosilicates. However, these two minerals share the structure of neither pyroxenes nor amphiboles, but are instead classified as members of the pyrorenoid group. Rhodonite is rose-red in colour due to the presence of manganese.

Phyllosilicates

The phyllosilicates or layer silicates contain sheetlike planes of $(SiO_4)^{4-}$ tetrahedra in which three of the four possible oxygen atoms are shared between different silicate tetrahedra. The term 'phyllosilicate' is derived from the Greek word phyllon, or 'leaf'; because of their sheeted structure the phyllosilicates tend to cleave into flakes or leaves. All members of the phyllosilicate group therefore have platy or flaky habit and display a single, prominent cleavage. Such minerals are soft and possess a low specific gravity.

Example members of the phyllosilicate class are kaolinite ($Al_4(Si_4O_{10})(OH)_4$), which is a major component of clay, serpentine ($Mg_6(Si_4O_{10})(OH)_8$), talc ($Mg_3(Si_4O_{10})(OH)_2$), and muscovite or white mica ($KAl_2(AlSi_3O_{10})(OH)_2$).

TECTOSILICATES

The tectosilicates or framework silicates are constructed in the form of a three-dimensional framework of $(SiO_4)^{4-}$ tetrahedra. In this structure all four of the oxygen ions belonging to each silicate tetrahedra are shared with nearby tetrahedra, forging a stable and strongly bonded network. The ratio of silicon to oxygen in this type of structure is 1:2.

The silica minerals are a subgroup of the tectosilicates. These include the nine different polymorphs of silicon dioxide, SiO_2: stishovite, coesite, quartz, keatite, cristobalite, tridymite, lechatelierite, and opal.

The feldspar group forms another subgroup of the tectosilicates. The feldspars are aluminium silicates which contain potassium (K), sodium (Na), or calcium (Ca) cations. Orthoclase ($K(AlSi_3O_8)$) and albite ($Na(AlSi_3O_8)$) offer two examples of feldspar minerals. Aluminium cations couple to four oxygen atoms to form a tetrahedral anion group, $(AlO_4)^{5-}$, which is structurally equivalent to that of the silicate radical $(SiO_4)^{4-}$. This aluminium anion group may then substitute for silicate radicals within the three-dimensional tectosilicate framework.

In order to maintain electrical neutrality during the substitution of $(AlO_4)^{5-}$ for $(SiO_4)^{4-}$, metal cations such as potassium, sodium, or calcium must be included. A third subgroup of the tectosilicates is composed of the zeolites, which are hydrous aluminium silicates. Natrolite ($Na_2(Al_2Si_3O_{10}).2H_2O$) and stilbite ($Ca(Al_2Si_7O_{18}).7H_2O$) are two examples of this group.

POLYMORPHISM IN SILICAS

There are nine different ways in which the silicate tetrahedra may be arranged in a continuous and electrically neutral structure in order to share all four oxygen ions. Therefore there exist nine different polymorphs of silicon dioxide, SiO_2. Each of these polymorphs possesses exactly the same chemical formula (SiO_2) as all the others, yet is built according to a different structural plan. The polymorphs adhere to different crystal systems and possess differing specific gravities. Which polymorph is stable in a particular environment is determined by the ambient temperature of the environment and the lattice energy of the polymorph. Those forms which are stable at higher temperatures possess more open, expanded lattices with greater lattice energies. The higher-temperature polymorphs of silicon dioxide therefore tend to display lower density and specific gravity.

The three major polymorphs of silicon dioxide are quartz, which possesses the most dense lattice and the lowest symmetry; tridymite, which has a less dense lattice and higher symmetry; and cristobalite, which has the most open lattice and the highest symmetry. A quantity of crystalline silicon dioxide may be transformed from one of these structural types to another only when the silicon-oxygen bonds are broken. The transition between cristobalite, tridymite and quartz thus takes place only over long periods of time or at extremes of temperature change. However, quartz, tridymite, and cristobalite also possess high- and low- temperature structural variations which are differentiated by shorter or longer bond lengths between the silicon and oxygen ions. Transitions between these high and low-temperature variations are fast and completely reversible.

2

Optical Mineralogy

Optical mineralogy is the study of minerals by measuring their optical properties. Most commonly, rock and mineral samples are prepared as thin sections or grain mounts for study in the laboratory with a petrographic microscope. Optical mineralogy is used to identify the mineralogical composition of geological materials in order to help reveal their origin and evolution.

Some of the properties and techniques used include:

Refractive Index

The refractive index (or index of refraction) of a medium is a measure for how much the speed of light (or other waves such as sound waves) is reduced inside the medium. For example, typical glass has a refractive index of 1.5, which means that light travels at 1 / 1.5 = 0.67 times the speed in air or vacuum.

Two common properties of glass and other transparent materials are directly related to their refractive index. First, light rays change direction when they cross the interface from air to the material, an effect that is used in lenses and glasses. Second, light reflects partially from surfaces that have a different refractive index than their surroundings.

Definition

The refractive index of a medium is defined as the ratio of the phase velocity of a wave phenomenon such as light or sound in a reference medium to the velocity in the medium itself.

It is most commonly used in the context of light with vacuum as a reference medium, although historically other reference media (e.g. air at a standardised pressure and temperature) have been common. It is usually given the symbol *n*. In the case of light, it equals

$$n = \sqrt{\epsilon_r}\sqrt{\mu_r},$$

where ε_r is the material's relative permittivity, and μ_r is its relative permeability. For most materials, μ_r is very close to 1 at optical frequencies, therefore n is approximately $\sqrt{\epsilon_r}$. Contrary to a widespread misconception, n may be less than 1. This has practical technical applications, such as effective mirrors for x-rays based on total internal reflection.

The phase velocity is defined as the rate at which the crests of the waveform propagate; that is, the rate at which the phase of the waveform is moving.

The *group velocity* is the rate that the *envelope* of the waveform is propagating; that is, the rate of variation of the amplitude of the waveform.

Provided the waveform is not distorted significantly during propagation, it is the group velocity that represents the rate that information (and energy) may be transmitted by the wave, for example the velocity at which a pulse of light travels down an optical fibre.

SPEED OF LIGHT

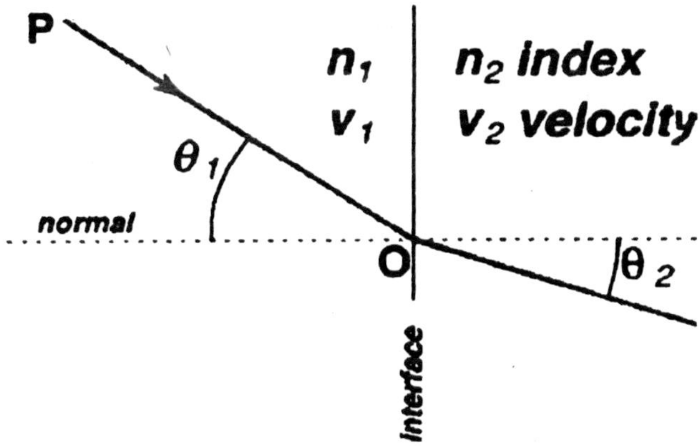

The speed of all electromagnetic radiation in vacuum is the same, approximately 3×10^8 metres per second, and is denoted by *c*. Therefore, if *v* is the phase velocity of radiation of a specific frequency in a specific material, the refractive index is given by

$$n = \frac{C}{v}$$

This number is typically greater than one: the higher the index of the material, the more the light is slowed down. However, at certain frequencies (e.g. near absorption resonances, and for X-rays), *n* will actually be smaller than one. This does not contradict the theory of relativity, which holds that no information-carrying signal can ever propagate faster than *c*, because the phase velocity is not the same as the group velocity or the signal velocity.

Sometimes, a "group velocity refractive index", usually called the *group index* is defined:

$$n_g = \frac{C}{v_g}$$

where v_g is the group velocity. This value should not be confused with *n*, which is always defined with respect to the phase velocity. The group index can be written in terms of the wavelength dependence of the refractive index as

$$n_g = n - \lambda \frac{dn}{d\lambda},$$

where λ is the wavelength in vacuum. At the microscale, an electromagnetic wave's phase velocity is slowed in a material because the electric field creates a disturbance in the charges of each atom (primarily the electrons) proportional to the permittivity of the medium. The charges will, in general, oscillate slightly out of phase with respect to the driving electric field. The charges thus radiate their own electromagnetic wave that is at the same frequency but with a phase delay. The macroscopic sum of all such contributions in the material is a wave with the same frequency but shorter wavelength than the original, leading to a slowing of the wave's phase velocity. Most of the radiation from oscillating material charges will modify the incoming wave, changing its velocity. However, some net energy will be radiated in other directions.

If the refractive indices of two materials are known for a given frequency, then one can compute the angle by which radiation of that frequency will be refracted as it moves from the first into the second material from Snell's law.

If in a given region the values of refractive indices n or n_g were found to differ from unity (whether homogeniously, or isotropically, or not), then this region was distinct from vacuum in the above sense for lacking Poincare symmetry.

Negative Refractive Index

Recent research has also demonstrated the existence of negative refractive index which can occur if the real parts of both ε_r and μ_r are *simultaneously* negative, although that is a sufficient but not necessary condition. Not thought to occur naturally, this can be achieved with so called metamaterials and offers the possibility of perfect lenses and other exotic phenomena such as a reversal of Snell's law.

Dispersion and Absorption

In real materials, the polarisation does not respond instantaneously to an applied field. This causes dielectric loss, which can be expressed by a permittivity that is both complex and frequency dependent. Real materials are not perfect insulators either, i.e. they have non-zero direct current conductivity. Taking both aspects into consideration, we can define a complex index of refraction:

$$\tilde{n} = n - i\kappa$$

Here, n is the refractive index indicating the phase velocity as above, while κ is called the extinction coefficient, which indicates the amount of absorption loss when the electromagnetic wave propagates through the material. Both n and κ are dependent on the frequency (wavelength).

The effect that n varies with frequency (except in vacuum, where all frequencies travel at the same speed, c, is known as dispersion, and it is what causes a prism to divide white light into its constituent spectral colours, explains rainbows, and is the cause of chromatic aberration in lenses. In regions of the spectrum where the material does not absorb, the real part of the refractive index tends to increase with frequency. Near absorption peaks, the curve of the refractive index is a complex form given by the Kramers-Kronig relations, and can decrease with frequency.

Since the refractive index of a material varies with the frequency (and thus wavelength) of light, it is usual to specify the corresponding vacuum wavelength at which the refractive index is measured. Typically, this is done at various well-defined spectral emission lines; for example, n_D is the refractive index at the Fraunhofer "D" line, the centre of the yellow sodium double emission at 589.29 nm wavelength.

The Sellmeier equation is an empirical formula that works well in describing dispersion, and Sellmeier coefficients are often quoted instead of the refractive index in tables. For some representative refractive indices at different wavelengths.

As shown above, dielectric loss and non-zero DC conductivity in materials cause absorption. Good dielectric materials such as glass have extremely low DC conductivity, and at low frequencies the dielectric loss is also negligible, resulting in almost no absorption $(\kappa \approx 0)$. However, at higher frequencies (such as visible light), dielectric loss may increase absorption significantly, reducing the material's transparency to these frequencies.

The real and imaginary parts of the complex refractive index are related through use of the Kramers-Kronig relations. For example, one can determine a material's full complex refractive index as a function of wavelength from an absorption spectrum of the material.

Anisotropy

The refractive index of certain media may be different depending on the polarisation and direction of propagation of the light through the medium. This is known as birefringence or anisotropy and is described by the field of crystal optics. In the most general case, the *dielectric constant* is a rank-2 tensor (a 3 by 3 matrix), which cannot simply be described by refractive indices except for polarisations along principal axes.

In magneto-optic (gyro-magnetic) and optically active materials, the principal axes are complex (corresponding to elliptical polarisations), and the dielectric tensor is complex-Hermitian (for lossless media); such materials break time-reversal symmetry and are used, e.g. to construct Faraday isolators.

Non-linearity

The strong electric field of high intensity light (such as output of a laser) may cause a medium's refractive index to vary as the light passes through it, giving rise to non-linear optics. If the index varies quadratically with the field (linearly with the intensity), it is called the optical Kerr effect and causes phenomena such as self-focusing and self phase modulation. If the index varies linearly with the field (which is only possible in materials that do not possess inversion symmetry), it is known as the Pockels effect.

INHOMOGENEITY

If the refractive index of a medium is not constant, but varies gradually with position, the material is known as a gradient-index medium and is described by gradient index optics. Light travelling through such a medium can be bent or focused, and this effect can be exploited to produce lenses, some optical fibres and other devices. Some common mirages are caused by a spatially-varying refractive index of air.

APPLICATIONS

The refractive index of a material is the most important property of any optical system that uses refraction. It is used to calculate the focusing power of lenses, and the dispersive power of prisms.

Since refractive index is a fundamental physical property of a substance, it is often used to identify a particular substance, confirm its purity, or measure its concentration. Refractive index is used to measure solids (glasses and gemstones), liquids, and gases. Most commonly it is used to measure the concentration of a solute in an aqueous solution. A refractometer is the instrument used to measure refractive index. For a solution of sugar, the refractive index can be used to determine the sugar content.

BIREFRINGENCE

Birefringence, or double refraction, is the decomposition of a ray of light into two rays (the ordinary ray and the extraordinary ray) when it passes through certain types of material, such as calcite crystals, depending on the polarisation of the light. This effect can occur only if the structure of the material is anisotropic. If the material has a single axis of anisotropy, (i.e. it is uniaxial) birefringence can be formalised by assigning two different refractive indices to the material for different polarisations. The birefringence magnitude is then defined by:

$$\Delta n = n_e - n_o$$

where n_o and n_e are the refractive indices for polarisations perpendicular (ordinary) and parallel (extraordinary) to the axis of anisotropy respectively.

Birefringence can also arise in magnetic, not dielectric, materials, but substantial variations in magnetic permeability of materials are rare at optical frequencies.

Creating Birefringence

While birefringence is often found naturally (especially in crystals), there are several ways to create it in optically isotropic materials.

- Birefringence results when isotropic materials are deformed such that the isotropy is lost in one direction (i.e., stretched or bent). Example: Applying an electric field can induce molecules to line up or behave asymmetrically, introducing anisotropy and resulting in birefringence.

Applying a magnetic field can cause a material to be circularly birefringent, with different indices of refraction for oppositely-handed circular polarisations.

Examples of Birefringent Materials

Many plastics are birefringent, because their molecules are 'frozen' in a stretched conformation when the plastic is moulded or extruded. For example, cellophane is a cheap birefringent material. Birefringent materials are used in many devices which manipulate the polarisation of light, such as wave plates, polarising prisms, and Lyot filters.

There are many birefringent crystals: birefringence was

first described in calcite crystals by the Danish scientist Rasmus Bartholin in 1669.

Birefringence can be observed in amyloid plaque deposits such as are found in the brains of Alzheimer's victims. Modified proteins such as immunoglobulin light chains abnormally accumulate between cells, forming fibrils. Multiple folds of these fibres line up and take on a beta-pleated sheet conformation. Congo red dye intercalates between the folds and, when observed under polarised light, causes birefringence.

Cotton (Gossypium hirsutum) fibre is birefringent because of high levels of cellulosic material in the fibre's secondary cell wall.

Slight imperfections in optical fibre can cause birefringence, which can cause distortion in fibre-optic communication.

Silicon carbide, also known as Moissanite, is strongly birefringent.

The refractive indices of several (uniaxial) birefringent materials are listed below (at wavelength ~ 590 nm), from.

Material	n_o	n_e	Δn
beryl $Be_3Al_2(SiO_3)6$	1.602	1.557	-0.045
calcite $CaCO_3$	1.658	1.486	-0.172
calomel Hg_2Cl_2	1.973	2.656	+0.683
ice H_2O	1.309	1.313	+0.014
lithium niobate $LiNbO_3$	2.272	2.187	-0.085
magnesium fluoride MgF_2	1.380	1.385	+0.006
quartz SiO_2	1.544	1.553	+0.009
ruby Al_2O_3	1.770	1.762	-0.008
rutile TiO_2	2.616	2.903	+0.287
peridot (Mg, Fe)2SiO_4	1.690	1.654	-0.036
sapphire Al_2O_3	1.768	1.760	-0.008
sodium nitrate $NaNO_3$	1.587	1.336	-0.251
tourmaline (complex silicate)	1.669	1.638	-0.031
zircon, high $ZrSiO_4$	1.960	2.015	+0.055
zircon, low $ZrSiO_4$	1.920	1.967	+0.047

Biaxial Birefringence

Biaxial birefringence, also known as trirefringence, describes an anisotropic material that has more than one axis of anisotropy. For such a material, the refractive index tensor n, will in general have three distinct eigenvalues that can be labelled n_α, n_β and n_γ.

The refractive indices of some trirefringent materials are listed below (at wavelength ~ 590 nm), from.

Material	n_α	n_β	n_γ
borax	1.447	1.469	1.472
epsom salt $MgSO_4 \cdot 7(H_2O)$	1.433	1.455	1.461
mica, biotite	1.595	1.640	1.640
mica, muscovite	1.563	1.596	1.601
olivine $(Mg, Fe)_2SiO_4$	1.640	1.660	1.680
perovskite $CaTiO_3$	2.300	2.340	2.380
topaz	1.618	1.620	1.627
ulexite	1.490	1.510	1.520

Measuring Birefringence

Birefringence and related optical effects (such as optical rotation and linear or circular dichroism) can be measured by measuring the changes in the polarisation of light passing through the material. These measurements are known as polarimetry.

A common feature of optical microscopes is a pair of crossed polarising filters. Between the crossed polarisers, a birefringent sample will appear bright against a dark (isotropic) background.

Applications of Birefringence

Birefringence is widely used in optical devices, such as liquid crystal displays, light modulators, colour filters, wave plates, optical axis gratings, etc. It also plays important role in second harmonic generation and many other non-linear processes. It is also utilised in medical diagnostics. Needle biopsy of suspected gouty joints will be negatively birefringent if urate crystals are present.

Elastic Birefringence

Another form of birefringence is observed in anisotropic elastic materials. In these materials, shear waves split according to similar principles. The study of birefringent shear waves in the earth is a part of seismology. Birefringence is also used in optical mineralogy to determine the chemical composition, and history of minerals and rocks.

Pleochroism

Pleochroism is an optical phenomenon in which the stone appears to be different colours when observed at different angles.

Pleochroism is caused by the double refraction of light by a coloured gem or crystal. Light of different polarisations is bent different amounts by the crystal, and therefore follows different paths through the crystal. The components of a divided light beam follow different paths within the stone and travel at different speeds, and each path will absorb different colours of light. When the stone is observed at some angle, light following some combination of paths and polarisations will be present, each of which will have had light of different colours absorbed. At another angle, the light passing through the stone will be composed of another combination of light paths and polarisations, each with their own colour. The light passing through the stone will therefore have different colours when it is viewed from different angles, making the stone seem to be of different colours.

Some stones show two colours or shades and are called dichroic. Some show three and are trichroic. Gems are sometimes cut and set either to display pleochroism or to hide it, depending on the colours and their attractiveness.

Pleochroism is an extremely useful tool in mineralogy for

mineral identification, since minerals that are otherwise very similar often have very different pleochroic colour schemes. In such cases, a thin section of the mineral is used and examined under transmitted light in a microscope.

William Nicol, whose name is associated with the creation of the Nicol prism, seems to have been the first to prepare thin slices of mineral substances, and his methods were applied by Henry Thronton Maire Witham (1831) to the study of plant petrifactions. This method, of such far-reaching importance in petrology, was not at once made use of for the systematic investigation of rocks, and it was not until 1858 that Henry Clifton Sorby pointed out its value. Meanwhile the optical study of sections of crystals had been advanced by Sir David Brewster and other physicists and mineralogists and it only remained to apply their methods to the minerals visible in rock sections.

Sections

A good rock-section should be about one-thousandth of an inch in thickness, and is by no means very difficult to make. A thin splinter of the rock, about as large as a halfpenny may be taken; it should be as fresh as possible and free from obvious cracks. By grinding it on a plate of planed steel or cast iron with a little fine carborundum it is soon rendered flat on one side and is then transferred to a sheet of plate glass and smoothed with the very finest emery till all minute pits and roughnesses are removed and the surface is a uniform plane. The rock-chip is then washed, and placed on a copper or iron plate which is heated by a spirit or gas lamp. A microscopic glass slip is also warmed on this plate with a drop of viscous natural Canada balsam on its surface. The more volatile ingredients of the balsam are dispelled by the heat, and when that is accomplished the smooth, dry, warm rock is pressed firmly into contact with the glass plate so that the film of balsam intervening may be as thin as possible and free from air-bubbles. The preparation is allowed to cool and then the rock chip is again ground down as before, first with carborundum and, when it becomes transparent, with fine emery till the desired thickness is obtained. It is then cleaned, again heated with a little more balsam, and covered with a cover glass. The labour of grinding the first surface may be avoided by cutting off a smooth slice with an iron disk armed with crushed diamond powder. A second application of the slitter after the first face is smoothed and cemented to the glass will in expert hands leave a rock-section so thin as to be already transparent. In this way the preparation of a section may require only twenty minutes.

Microscope

The microscope employed is usually on which is provided with a rotating stage beneath which there is a polariser, while above the objective or the eyepiece an analyser is mounted; alternatively the stage may be fixed and the polarising and analysing prisms may be capable of simultaneous rotation by means of toothed wheels and a connecting-rod. If ordinary light and not polarised light is desired, both prisms may be withdrawn from the axis of the instrument; if the polariser only is inserted the light transmitted is plane polarised; with both prisms in position the slide is viewed between "crossed nicols." A microscopic rock-section in ordinary light if a suitable magnification (say 30) be employed is seen to consist of grains or crystals varying in colour, size and shape.

CHARACTERS OF MINERALS

Some minerals are colourless and transparent (quartz, calcite, felspar, muscovite, etc.), others are yellow or brown (rutile, tourmaline, biotite), green (diopside, hornblende, chlorite), blue (glaucophane), pink (garnet), etc. The same mineral may present a variety of colours, in the same or different rocks, and these colours may be arranged in zones parallel to the surfaces of the crystals.

Thus tourmaline may be brown, yellow, pink, blue, green,

violet, grey, or colourless, but every mineral has one or more characteristic, because most common tints. The shapes of the crystals determine in a general way the outlines of the sections of them presented on the slides. If the mineral has one or more good cleavages they will be indicated by systems of cracks.

The refractive index is also clearly shown by the appearance of the section, which are rough, with well-defined borders if they have a much stronger refraction than the medium in which they are mounted. Some minerals decompose readily and become turbid and semi-transparent (e.g. felspar); others remain always perfectly fresh and clear (e.g. quartz), others yield characteristic secondary products (such as green chlorite after biotite). The inclusions in the crystals are of great interest; one mineral may enclose another, or may contain spaces occupied by glass, by fluids or by gases.

Microstructure

Lastly the structure of the rock, that is to say, the relation of its components to one another, is usually clearly indicated, whether it be fragmented or massive; the presence of glassy matter in contradistinction to a completely crystalline or "holo-crystalline" condition; the nature and origin of organic fragments; banding, foliation or lamination; the pumiceous or porous structure of many lavas; these and many other characters, though often not visible in the hand specimens of a rock, are rendered obvious by the examination of a microscopic section.

Many refined methods of observation may be introduced, such as the measurement of the size of the elements of the rock by the help of micrometres; their relative proportions by means of a glass plate ruled in small squares; the angles between cleavages or faces seen in section by the use of the rotating graduated stage, and the estimation of the refractive index of the mineral by comparison with those of different mounting media.

Pleochroism

Pleochroism is an optical phenomenon in which the stone appears to be different colours when observed at different angles.

Pleochroism is caused by the double refraction of light by a coloured gem or crystal. Light of different polarisations is bent different amounts by the crystal, and therefore follows different paths through the crystal. The components of a divided light beam follow different paths within the stone and travel at different speeds, and each path will absorb different colours of light. When the stone is observed at some angle, light following some combination of paths and polarisations will be present, each of which will have had light of different colours absorbed. At another angle, the light passing through the stone will be composed of another combination of light paths and polarisations, each with their own colour. The light passing through the stone will therefore have different colours when it is viewed from different angles, making the stone seem to be of different colours.

Some stones show two colours or shades and are called dichroic. Some show three and are trichroic. Gems are sometimes cut and set either to display pleochroism or to hide it, depending on the colours and their attractiveness.

Pleochroism is an extremely useful tool in mineralogy for mineral identification, since minerals that are otherwise very similar often have very different pleochroic colour schemes. In such cases, a thin section of the mineral is used and examined under transmitted light in a microscope.

Double Refraction

If the analyser be now inserted in such a position that it is crossed relatively to the polariser the field of view will be dark where there are no minerals, or where the light passes through isotropic substances such as glass, liquids and cubic crystals. All other crystalline bodies, being doubly refracting, will appear bright in some position as the stage is rotated. The only exception to this rule is provided by sections which are perpendicular to the optic axes of birefringent crystals; these remain dark or nearly dark during a whole rotation, and as will be seen later, their investigation is of special importance.

Extinction

The doubly refracting mineral sections, however, will in all cases appear black in certain positions as the stage is rotated. They are said to go "extinct" when this takes place. If we note these positions we may measure the angle between them and any cleavages, faces or other structures of the crystal by means of the rotating stage. These angles are characteristic of the system to which the mineral belongs and often of the mineral species itself. To facilitate measurement of extinction angles various kinds of eyepieces have been devised, some having a stereoscopic calcite plate, others with two or four plates of quartz cemented together; these are often found to give more exact results than are obtained by observing merely the position in which the mineral section is most completely dark between crossed nicols.

The mineral sections when not extinguished are not only bright but are coloured and the colours they show depend on several factors, the most important of which is the strength of the double refraction. If all the sections are of the same thickness as is nearly true of well-made slides, the minerals with strongest double refraction yield the highest polarisation colours. The order in which the colours are arranged in what is known as Newton's scale, the lowest being dark grey, then grey, white, yellow, orange, red, purple, blue and so on. The difference between the refractive indexes of the ordinary and the extraordinary ray in quartz is .009, and in a rock-section about 1/500 of an inch thick this mineral gives grey and white polarisation things; nepheline with weaker double refraction gives dark grey; augite on the other hand will give red and blue, while calcite with the stronger double refraction will appear pinkish or greenish white. All sections of the same mineral, however, will not have the same colour; it was stated above that sections perpendicular to an optic axis will be nearly black, and, in general, the more nearly any section approaches this direction the lower its polarisation colours will be.

By taking the average, or the highest colour given by any mineral, the relative value of its double refraction can be estimated; or if the thickness of the section be precisely known the difference between the two refractive indexes can be ascertained. If the slides be thick the colours will be on the whole higher than in thin slides.

It is often important to find out whether of the two axes of elasticity (or vibration traces) in the section is that of greater elasticity (or lesser refractive index). The quartz wedge or selenite plate enables us to do this. Suppose a doubly refracting mineral section so placed that it is "extinguished"; if now is rotated through 45 degrees it will be brightly illuminated. If the quartz wedge be passed across it so that the long axis of the wedge is parallel to the axis of elasticity in the section the polarisation colours will rise or fall. If they rise the axes of greater elasticity in the two minerals are parallel; if they sink the axis of greater elasticity in the one is parallel to that of lesser elasticity in the other. In the latter case by pushing the wedge sufficiently far complete darkness or compensation will result. Selenite wedges, selenite plates, mica wedges and mica plates are also used for this purpose.

A quartz wedge also may be calibrated by determining the amount of double refraction in all parts of its length. If now it be used to produce compensation or complete extinction in any doubly refracting mineral section, we can ascertain what is the strength of the double refraction of the section because it is obviously equal and opposite to that of a known part of the quartz wedge.

A further refinement of microscopic methods consists of the use of strongly convergent polarised light (konoscopic methods). This is obtained by a wide angled achromatic condenser above the polariser, and a high power microscopic objective. Those sections are most useful which are perpendicular to an optic axis, and consequently remain dark on rotation. If they belong to uniaxial crystals they show a dark cross or convergent light between crossed nicols, the bars of which remain parallel to the wires in the field of the eyepiece. Sections perpendicular to an optic axis of a biaxial mineral under the same conditions show a dark bar which on rotation becomes curved to a hyperbolic shape. If the section is perpendicular to a "bisectrix" a black cross is seen which on rotation opens out to form two hyperbolas, the apices of which are turned towards one another.

The optic axes emerge at the apices of the hyperbolas and may be surrounded by coloured rings, though owing to the thinness of minerals in rock sections these are only seen when the double refraction of the mineral is strong. The distance between the axes as seen in the field of the microscope depends partly on the axial angle of the crystal and partly on the numerical aperture of the objective. If it is measured by means of eye-piece micrometre, the optic axial angle of the mineral can be found by a simple calculation. The quartz wedge, quarter mica plate or selenite plate permit the determination of the positive or negative character of the crystal by the changes in the colour or shape of the figures observed in the field. These operations are precisely similar to those employed by the mineralogist in the examination of plates cut from crystals. It is sufficient to point out that the petrological microscope in its modern development is an optical instrument of great precision, enabling us to determine physical constants of crystallised substances as well as serving to produce magnified images like the ordinary microscope. A great variety of accessory apparatus has been devised to fit it for these special uses.

Examination of Rock Powders

Although rocks are now studied principally in microscopic sections the investigation of fine crushed rock powders, which was the first branch of microscopic petrology to receive attention, is by no means discontinued. The modern optical methods are perfectly applicable to transparent mineral fragments of any kind.

Minerals are almost as easily determined in powder as in section, but it is otherwise with rocks, as the structure or relation of the components to one another, which is an element of great importance in the study of the history and classification or rocks, is almost completely destroyed by grinding them to powder.

3

Gemology

Gemology is the science, art and profession of identifying and evaluating gemstones. It is considered a geo-science and a branch of mineralogy. Some jewellers are academically trained gemologists and are qualified to identify and evaluate gems.

With further post-graduate training and experience they can qualify as appraisers (valuers). There are numerous professional schools and associations of gemologists and certification programmes all over the world.

Some gemologists specialise in various gemstones, such as diamonds or emeralds. The laboratory gemologists are expert in identification of origin.

Recently, the demand for gemological services has grown, as increasing quantities of synthetic gems such as cubic zirconia and synthetic moissanite are manufactured. Gemologists perform such work as the identification of synthetic and natural gemstones, fracture-filled gemstones, and colour-enhanced or treated natural gemstones.

Gemstones are basically categorised based on of their crystal structure, specific gravity, refractive index, and other optical properties, such as pleochroism. The physical property of "hardness" is defined by the non-linear Mohs scale of mineral hardness.

Gemologists study these factors while valuing or appraising cut and polished gemstones. Gemological microscopic study of the internal structure is used to determine whether a gem is synthetic or natural by revealing natural fluid inclusions, and included partially melted exogenous crystals to demonstrate evidence of heat treatment to enhance colour.

The spectroscopic analysis of cut gemstones also allows a gemologist to understand the atomic structure and identify its origin as it is a major factor in valuing a gemstone.

For example, a ruby from Burma will have definite internal and optical activity variance as compared to a Thai ruby.

When the gemstones are in a rough state, the gemologist studies the external structure; the host rock and mineral association; and natural and polished colour. Initially, the stone is identified by its colour, refractive index, optical character, specific gravity, and examination of internal characteristics under magnification.

Gemstones

Diamond: Diamond is the hardest natural material known to man and the third-hardest known material after aggregated diamond nanorods and ultrahard fullerite. Its hardness and high dispersion of light make it useful for industrial applications and jewellery.

Diamonds are specifically renowned as a material with superlative physical qualities — they make excellent abrasives because they can be scratched only by other diamonds, Borazon, ultrahard fullerite, or aggregated diamond nanorods, which also means they hold a polish extremely well and retain their lustre. About 130 million carats (26,000 kg) are mined annually, with a total value of nearly USD $9 billion. About 100,000 kg are synthesised annually.

The name diamond derives from the ancient Greek *adamas* "invincible". They have been treasured as gemstones since their use as religious icons in ancient India and usage in engraving tools also dates to early human history. Popularity of diamonds has risen since the 19th century because of increased supply, improved cutting and polishing techniques, growth in the world economy, and innovative and successful advertising campaigns. They are commonly judged by the "four Cs": carat, clarity, colour, and cut.

Roughly 49 per cent of diamonds originate from central and southern Africa, although significant sources of the mineral have been discovered in Canada, India, Russia, Brazil, and Australia. They are mined from kimberlite and lamproite volcanic pipes, which brought to the surface the diamond crystals from deep in the Earth where the high pressure and temperature enables the formation of the crystals.

The mining and distribution of natural diamonds are subjects of frequent controversy such as with concerns over the sale of conflict diamonds by African paramilitary groups. There are also allegations that the De Beers Group misuses its dominance in the industry to control supply and manipulate price via monopolistic practices, although in recent years the company's market share has dropped to below 50 per cent.

Material Properties

A diamond is a transparent crystal of tetrahedrally bonded carbon atoms. Diamonds have been adapted for many uses because of the material's exceptional physical characteristics. Most notable are its extreme hardness, its high dispersion index, and high thermal conductivity, with a melting point of 3820 K, which converts to 6416.33 F and 3546.85 C.

Hardness

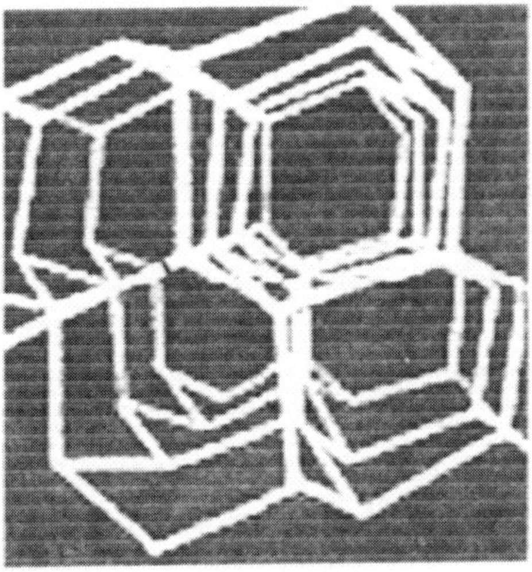

The atomic connectivity of the carbon atoms gives the gem its hardness

Diamond is the hardest natural material that is known, its hardness set to 10, i.e. hardest, on Mohs scale of mineral hardness and having an absolute hardness value of between 90, 167, and 231 gigapascals in various tests. Diamond's hardness has been known since antiquity, and is the source of its name. However, aggregated diamond nanorods, an allotrope of carbon first synthesised in 2005, are even harder than diamond.

The hardest diamonds in the world are from the New England area in New South Wales, Australia. These diamonds are generally small, perfect to semi-perfect octahedra, and are used to polish other diamonds. Their hardness is considered to be a product of the crystal growth form, which is single stage growth crystal. Most other diamonds show more evidence of multiple growth stages, which produce inclusions, flaws, and defect planes in the crystal lattice all of which affect their hardness.

The hardness of diamonds contributes to its suitability as a gemstone. Because it can only be scratched by other diamonds, it maintains its polish extremely well, keeping its lustre over long periods of time. Unlike many other gems, it is well-suited to daily wear because of its resistance to scratching—perhaps contributing to its popularity as the preferred gem in an engagement ring or wedding ring, which are often worn every day.

Industrial use of diamonds has historically been associated with their hardness; this property makes diamond the ideal material for cutting and grinding tools. As the hardest known naturally occurring material, diamond can be used to polish, cut, or wear away any material, including other diamonds. Common industrial adaptations of this ability include diamond-tipped drill bits and saws, or use of diamond powder as an abrasive. Industrial-grade diamonds are either unsuitable for use as gems or synthetically produced, which lowers their value and makes their use economically feasible. Industrial applications, especially as drill bits and engraving tools, also date to ancient times.

Electrical Conductivity

Other specialised applications also exist or are being developed, including use as semi-conductors: some blue diamonds are natural semi-conductors, in contrast to most other diamonds, which are excellent electrical insulators.

Toughness

Toughness relates to a material's ability to resist breakage from forceful impact. The toughness of natural diamond has been measured as 3.4 MN $m^{-3/2}$, which is good compared to other gemstones, but poor compared to most engineering materials. As with any material, the macroscopic geometry of a diamond contributes to its resistance to breakage. Diamond is therefore more fragile in some orientations than others.

Colour

Diamonds can occur in nearly any colour, though yellow

and brown are by far the most common. "Black" diamonds are not truly black, but rather contain numerous dark inclusions that give the gems their dark appearance. When the colour is saturated enough in yellow or brown diamonds, a stone may be referred to as a *fancy coloured diamond* by the gem trade, otherwise they are graded for colour in the *normal colour range* of white diamonds. Coloured diamonds contain impurities or structural defects that cause the colouration, while pure or nearly pure diamonds are transparent and colourless. Most diamond impurities replace a carbon atom in the crystal lattice, known as a carbon flaw.

The most common impurity, nitrogen, causes a slight to intense yellow colouration depending upon the type and concentration of nitrogen present. The Gemological Institute of America (GIA) classifies low saturation yellow and brown diamonds as diamonds in the *normal colour range*, and applies a grading scale from 'D' (colourless) to 'Z' (light yellow). The GIA labels diamonds that have more colour than a 'Z' diamond *fancy*, along with those that are any colour other than yellow or brown.

Natural History

Formation

The formation of natural diamond requires very specific conditions. Diamond formation requires exposure of carbon-bearing materials to high pressure, ranging approximately between 45 and 60 kilobars, but at a comparatively low temperature range between approximately 1652-2372F (900-1300C). These conditions are known to be met in two places on Earth; in the lithospheric mantle below relatively stable continental plates, and at the site of a meteorite strike.

Diamonds Formed in Cratons

The conditions required for diamond formation in the lithospheric mantle require considerable depth. The range of depth required for diamond formation is calculated to be between 90-120 miles (140-190 kilometres). The rate at which temperature changes with increasing depth into the Earth (geothermal gradient) varies greatly in different parts of the Earth. In particular, under oceanic plates the temperature rises more quickly with depth, beyond the range required for diamond formation at the depth required. The correct combination of temperature and pressure is only found in the thick, ancient, and stable parts of continental plates where regions of lithosphere known as *cratons* exist. Long residence in the cratonic lithosphere allows diamond crystals to grow larger.

Through studies of carbon isotope ratios (similar to the methodology used in carbon dating, except with the stable isotopes C-12 and C-13), it has been shown that the carbon found in diamonds comes from both inorganic and organic sources. Some diamonds, known as *harzburgitic*, are formed from inorganic carbon originally found deep in the Earth's mantle.

In contrast, *eclogitic* diamonds contain organic carbon from organic detritus that has been pushed down from the surface of the Earth's crust through subduction before transforming into diamond. These two different source carbons have measurably different $^{13}C:^{12}C$ ratios. Diamonds that have come to the Earth's surface are generally very old, ranging from under 1 billion to 3.3 billion years old.

Diamonds occur most often as euhedral or rounded octahedra and twinned octahedra known as *macles* or *maccles*. As diamond's crystal structure has a cubic arrangement of the atoms, they have many facets that belong to a cube, octahedron, rhombicosidodecahedron, tetrakis hexahedron or disdyakis dodecahedron. The crystals can have rounded off and unexpressive edges and can be elongated. Sometimes they are found grown together or form double "twinned" crystals grown together at the surfaces of the octahedron. This is all due to the conditions in which they form. Diamonds (especially those with rounded crystal faces) are commonly found coated in *nyf,* an opaque gum-like skin.

Diamonds and Meteorite Strikes

Diamonds can also form in other natural high-pressure, relatively low-temperature events. Very small diamonds, known as *microdiamonds* or *nanodiamonds,* have been found in impact craters where meteors strike the Earth and create shock zones of high pressure and temperature where diamond formation can occur. Microdiamonds are now used as one indicator of ancient meteorite impact sites.

Surfacing

Diamond-bearing rock is brought close to the surface through deep-origin volcanic eruptions. The magma for such a volcano must originate at a depth where diamonds can be formed, 150 km (90 miles) deep or more (three times or more the depth of source magma for most volcanoes); this is a relatively rare occurrence. These typically small surface volcanic craters extend downward in formations known as volcanic pipes. The pipes contain material that was transported towards the surface by volcanic action, but was not ejected before the volcanic activity ceased.

During eruption these pipes are open to the surface, resulting in open circulation; many xenoliths of surface rock and even wood and/or fossils are found in volcanic pipes. Diamond-bearing volcanic pipes are closely related to the oldest, coolest regions of continental crust (cratons). This is because cratons are very thick, and their lithospheric mantle extends to great enough depth that diamonds are stable. Not all pipes contain diamonds, and even fewer contain enough diamonds to make mining economically viable.

The magma in volcanic pipes is usually one of two characteristic types, which cool into igneous rock known as either kimberlite or lamproite. The magma itself does not contain diamond; instead, it acts as an elevator that carries deep-formed rocks (xenoliths), minerals (xenocrysts), and fluids upward. These rocks are characteristically rich in magnesium-bearing olivine, pyroxene, and amphibole minerals which are often altered to serpentine by heat and fluids during and after eruption. Certain *indicator minerals* typically occur within diamondiferous kimberlites and are used as mineralogic tracers by prospectors, who follow the indicator trail back to the volcanic pipe which may contain diamonds. These minerals are rich in chromium (Cr) or titanium (Ti), elements which impart bright colours to the minerals.

The most common indicator minerals are chromium garnets (usually bright red Cr-pyrope, and occasionally green ugrandite-series garnets), eclogitic garnets, orange Ti-pyrope, red high-Cr spinels, dark chromite, bright green Cr-diopside, glassy green olivine, black picroilmenite, and magnetite. Kimberlite deposits are known as *blue ground* for the deeper serpentinised part of the deposits, or as *yellow ground* for the near surface smectite clay and carbonate weathered and oxidised portion.

Once diamonds have been transported to the surface by magma in a volcanic pipe, they may erode out and be distributed over a large area. A volcanic pipe containing diamonds is known as a *primary source* of diamonds. *Secondary sources* of diamonds include all areas where a significant number of diamonds, eroded out of their kimberlite or lamproite matrix, accumulate because of water or wind action. These include alluvial deposits and deposits along existing and ancient shorelines, where loose diamonds tend to accumulate because of their approximate size and density. Diamonds have also rarely been found in deposits left behind by glaciers (notably in Wisconsin and Indiana); however, in contrast to alluvial deposits, glacial deposits are not known to be of significant concentration and are therefore not viable commercial sources of diamond.

Diamonds can also be brought to the surface through certain processes which may occur when two continental plates collide and deeply formed rock is thrust to the surface, although this phenomenon is less understood and currently assumed to be uncommon.

Gemological Characteristics

The most familiar usage of diamonds today is as gemstones used for adornment. This usage dates back into antiquity and predates other uses. The dispersion of white light into spectral colours, is the primary gemological characteristic of gem diamonds. In the twentieth century, experts in the field of *gemology* have developed methods of grading diamonds and other gemstones based on the characteristics most important to their value as a gem. Four characteristics, known informally as the *four Cs,* are now commonly used as the basic descriptors of diamonds: these are *carat, cut, colour,* and *clarity.*

Most gem diamonds are traded on the wholesale market based on single values for each of the four Cs; for example knowing that a diamond is rated as 1.5 carats, VS2 clarity, F colour, excellent cut round brilliant, is enough to reasonably establish an expected price range. More detailed information from within each characteristic is used to determine actual market value for individual stones. Consumers who purchase individual diamonds are often advised to use the four Cs to pick the diamond that is "right" for them.

Other characteristics not described by the four Cs influence the value or appearance of a gem diamond. These characteristics include physical characteristics such as the presence of fluorescence, as well as data on a diamond's history including its source and which gemological institute performed evaluation services on the diamond. *Cleanliness* also dramatically affects a diamond's beauty.

There are three major non-profit gemological associations which grade and provide reports on diamonds. While carat weight and cut angles are mathematically defined, the clarity and colour are judged by the trained human eye and are therefore open to slight variance in interpretation.

Gemological Institute of America (GIA) was the first laboratory to issue modern diamond reports, and is held in high regard amongst gemologists for its consistent, conservative grading.

- American Gemological Society (AGS) is not as widely recognised nor as old as the GIA but garners a high reputation. The AGS employs a number system for grading cut quality, colour grade, and clarity. The highest grade being '0', and the lowest being '10'.
- Diamond High Council (HRD) Official certification laboratory of the Belgian diamond industry is located in Antwerp.
- Diamond Certification Laboratory of Australia—The DCLA is one of only five laboratories worldwide recognised as an International Diamond Council (IDC) laboratory. DCLA is the official CIBJO laboratory for Australia and holds the official registered National Diamond Masterset for Australia. The DCLA is the laboratory endorsed by the Jewellers Association of Australia.

Carat

The *carat weight* measures the mass of a diamond. One carat is defined as a fifth of a gram, or exactly 200 milligrams (about 0.007 ounce avoirdupois). The *point* unit—equal to one-hundredth of a carat (0.01 carat, or 2 mg)—is commonly used for diamonds of less than one carat. All else being equal, the value of a diamond increases exponentially in relation to carat weight, since larger diamonds are both rarer and more desirable for use as gemstones. A review of comparable diamonds available for purchase in September 2005 demonstrates this effect (approximate prices for round cut, G colour, VS2 diamonds with "1A" cut grade.

Carat size	*Cost per carat (US$)*	*Total cost (US$)*
0.5 carat (50 points)	3,000	1,500
1.0 carat	6,500	6,500
1.5 carats	8,500	12,750
2.0 carats	13,000	26,000
3.0 carats	17,000	51,000
5.0 carats	23,000	115,000

The price per carat does not increase smoothly with increasing size. Instead, there are sharp jumps around milestone carat weights, as demand is much higher for diamonds weighing just more than a milestone than for those weighing just less. As an example, a 0.95 carat diamond may have a significantly lower price per carat than a comparable 1.05 carat diamond, because of differences in demand.

A weekly diamond price list, the Rapaport Diamond Report is published by Martin Rapaport, CEO of Rapaport Group of New York, for different diamond cuts, clarity and weights. It is currently considered the de-facto retail price baseline. Jewellers often trade diamonds at negotiated discounts off the Rapaport price (e.g., "R -3 per cent").

In the wholesale trade of gem diamonds, carat is often used in denominating lots of diamonds for sale. For example, a buyer may place an order for 100 carats of 0.5 carat, D–F, VS2-SI1, excellent cut diamonds, indicating he wishes to purchase 200 diamonds (100 carats total mass) of those approximate characteristics. Because of this, diamond prices (particularly among wholesalers and other industry professionals) are often quoted per carat, rather than per stone.

Total carat weight (t.c.w.) is a phrase used to describe the total mass of diamonds or other gemstone in a piece of jewellery, when more than one gemstone is used. Diamond solitaire earrings, for example, are usually quoted in t.c.w. when placed for sale, indicating the mass of the diamonds in both earrings and not each individual diamond. T.c.w. is also widely used for diamond necklaces, bracelets and other similar jewellery pieces.

Clarity

Clarity is a measure of internal defects of a diamond called *inclusions*. Inclusions may be crystals of a foreign material or another diamond crystal, or structural imperfections such as tiny cracks that can appear whitish or cloudy. The number, size, colour, relative location, orientation, and visibility of inclusions can all affect the relative clarity of a diamond. The Gemological Institute of America (GIA) and other organisations have developed systems to grade clarity, which are based on those inclusions which are visible to a trained professional when a diamond is viewed under 10x magnification.

Diamonds become increasingly rare when considering higher clarity gradings. Only about 20 per cent of all diamonds mined have a clarity rating high enough for the diamond to be considered appropriate for use as a gemstone; the other 80 per cent are relegated to industrial use. Of that top 20 per cent, a significant portion contains one or more visible inclusions. Those that do not have a visible inclusion are known as "eye-clean" and are preferred by most buyers, although visible inclusions can sometimes be hidden under the setting in a piece of jewellery.

Most inclusions present in gem-quality diamonds do not affect the diamonds' performance or structural integrity. However, large clouds can affect a diamond's ability to transmit and scatter light. Large cracks close to or breaking the surface may reduce a diamond's resistance to fracture.

Diamonds are graded by the major societies on a scale ranging from flawless to imperfect.

Colour

A chemically pure and structurally perfect diamond is perfectly transparent with no hue, or colour. However, in reality almost no gem-sized natural diamonds are absolutely perfect. The colour of a diamond may be affected by chemical impurities and/or structural defects in the crystal lattice.

Depending on the hue and intensity of a diamond's colouration, a diamond's colour can either detract from or enhance its value. For example, most white diamonds are discounted in price as more yellow hue is detectable, while intense pink or blue diamonds (such as the Hope Diamond) can be dramatically more valuable. The Aurora Diamond Collection displays a spectacular array of naturally coloured diamonds, and is a good source of information on the topic of coloured diamonds.

Most diamonds used as gemstones are basically transparent with little tint, or *white diamonds*. The most common impurity, nitrogen, replaces a small proportion of carbon atoms in a diamond's structure and causes a yellowish to brownish tint. This effect is present in almost all white diamonds; in only the rarest diamonds is the colouration due to this effect undetectable. The GIA has developed a rating system for colour in white diamonds, from "D" to "Z" (with D being "colourless" and Z having a bright yellow colouration), which has been widely adopted in the industry and is universally recognised, superseding several older systems once used in different countries. The GIA system uses a benchmark set of natural diamonds of known colour grade, along with standardised and carefully controlled lighting conditions. Precision-crafted cubic zirconia master sets are sometimes used in the trade, however the GIA has found these sets to be inaccurate.

Diamonds with higher colour grades are rarer, in higher demand, and therefore more expensive, than lower colour grades. Oddly enough, diamonds graded Z are also rare, and the bright yellow colour is also highly valued. Diamonds graded D-F are considered "colourless", G-J are considered "near-colourless", K-M are "slightly coloured". N-Y usually appear light yellow or brown.

In contrast to yellow or brown hues, diamonds of other colours are more rare and valuable. While even a pale pink or blue hue may increase the value of a diamond, more intense colouration is usually considered more desirable and commands the highest prices. A variety of impurities and structural imperfections cause different colours in diamonds, including yellow, pink, blue, red, green, brown, and other hues. Diamonds with unusual or intense colouration are sometimes libelled "fancy" by the diamond industry. Intense yellow colouration is considered one of the fancy colours, and is separate from the colour grades of white diamonds. Gemologists have developed rating systems for fancy coloured diamonds, but they are not in common use because of the relative rarity of coloured diamonds.

Cut

Diamond cutting is the art and science of creating a gem-quality diamond out of mined rough. The *cut* of a diamond describes the manner in which a diamond has been shaped and polished from its beginning form as a rough stone to its final gem proportions. The cut of a diamond describes the quality of workmanship and the angles to which a diamond is cut. Often diamond cut is confused with "shape".

There are mathematical guidelines for the angles and length ratios at which the diamond is supposed to be cut in order to reflect the maximum amount of light. Round brilliant diamonds, the most common, are guided by these specific guidelines, though fancy cut stones are not able to be as accurately guided by mathematical specifics.

The techniques for cutting diamonds have been developed over hundreds of years, with perhaps the greatest achievements made in 1919 by mathematician and gem enthusiast Marcel Tolkowsky. He developed the round brilliant cut by calculating the ideal shape to return and scatter light when a diamond is viewed from above. The modern round brilliant has 57 facets (polished faces), counting 33 on the *crown* (the top half), and 24 on the *pavilion* (the lower half). The girdle is the thin middle part. The function of the crown is to diffuse light into various colours and the pavilion's function to reflect light back through the top of the diamond.

Tolkowsky defines the ideal dimensions to have:

- Table percentage (table diameter divided by overall diameter) = 53 per cent
- Depth percentage (Overall depth divided by the overall diameter) = 59.3 per cent
- Pavilion Angle (Angle between the girdle and the pavilion) = 40.75°
- Crown Angle (Angle between the girdle and the crown) = 34.5°
- Pavilion Depth (Depth of pavilion divided by overall diameter) = 43.1 per cent
- Crown Depth (Depth of crown divided by crown diameter) = 16.2 per cent

The culet is the tiny point or facet at the bottom of the diamond. This should be a negligible diameter, otherwise light leaks out of the bottom. Tolkowsky's ideal dimensions did not include a girdle. However, a thin girdle is required in reality in order to prevent the diamond from easily chipping in the setting. A normal girdle should be about 1 per cent–2 per cent of the overall diameter.

The further the diamond's characteristics are from Tolkowsky's ideal, the less light will be reflected. However, there is a small range in which the diamond can be considered "ideal." Today, because of the relative importance of carat weight in society, many diamonds are often intentionally cut poorly to increase carat weight. There is a financial premium for a diamond that weighs the magical 1.0 carat, so often the girdle is made thicker or the depth is increased. Neither of these tactics make the diamond appear any bigger, and they greatly reduce the sparkle of the diamond. So a poorly cut 1.0 carat diamond may have the same diameter and appear as large as a 0.85 carat diamond. The depth percentage is the overall quickest indication of the quality of the cut of a round brilliant. "Ideal" round brilliant diamonds should not have a depth percentage greater than 62.5 per cent. Another quick indication is the overall diameter. Typically a round brilliant 1.0 carat diamond should have a diameter of about 6.5 mm. Mathematically, the diameter in millimetres of a round brilliant should approximately equal 6.5 times the cube root of carat weight, or 11.1 times the cube root of gram weight, or 1.4 times the cube root of point weight.

Ideal cuts can be controversial as the definitions of brilliance and beauty are very subjective.

Tolkowsky's mathematical model is now superseded by the GIA Facetware software that is the culmination of 20 years of studies on diamond cuts.

New diamond cuts are now all the rage in the diamond industry as for example a design invented in 2003 and called the Genesis cut. This cut differs in shape from the more traditional cuts in its concave surfaces and angles and resembles a 4-pointed star.

Shape

Diamonds do not show all of their beauty as rough stones; instead, they must be cut and polished to exhibit the characteristic fire and brilliance that diamond gemstones are known for. Diamonds are cut into a variety of shapes that are generally designed to accentuate these features.

Diamonds which are not cut to the specifications of Tolkowsky's round brilliant shape (or subsequent variations) are known as "fancy cuts." Popular fancy cuts include the *baguette* (from the French, meaning *rod* or loaf of bread), *marquise, princess cut* (square outline), *heart, briolette* (a form of the rose cut), and *pear* cuts. Newer cuts that have been introduced into the jewellery industry are the "cushion" "radiant"(similar to princess cuts, but with rounded edges instead of square edges) and "Asscher" cuts. Many fancy coloured diamonds are now being cut according to these new styles. Generally speaking, these "fancy cuts" are not held to the same strict standards as Tolkowsky-derived round brilliants and there are less specific mathematical guidelines of angles which determine a well-cut stone. Cuts are influenced heavily by fashion: the baguette cut—which accentuates a diamond's lustre and downplays its fire—was all the rage during the Art Deco period, whereas the princess cut—which accentuates a diamond's fire rather than its lustre—is currently gaining popularity. The princess cut is also popular amongst diamond cutters: of all the cuts, it wastes the least of the original crystal. The past decades have seen the development of new diamond cuts, often based on a modification of an existing cut. Some of these include extra facets. These newly developed cuts are viewed by many as more of an attempt at brand differentiation by diamond sellers, than actual improvements to the state of the art.

Quality

The quality of a diamond's cut is widely considered the most important of the four Cs in determining the beauty of a diamond; indeed, it is commonly acknowledged that a well-cut diamond can appear to be of greater carat weight, and have clarity and colour appear to be of better grade than they actually are. The skill with which a diamond is cut determines its ability to reflect and refract light.

In addition to carrying the most importance to a diamond's quality as a gemstone, the cut is also the most difficult to quantitatively judge. A number of factors, including proportion, polish, symmetry, and the relative angles of various facets, are determined by the quality of the cut and can affect the performance of a diamond. A poorly cut diamond with facets cut only a few degrees out of alignment can result in a poorly performing stone. For a round brilliant cut, there is a balance between "brilliance" and "fire." When a diamond is cut for too much "fire," it looks like a cubic zirconia, which gives off much more "fire" than real diamond. A well-executed round brilliant cut should reflect light upwards and make the diamond appear white when viewed from the top. An inferior cut will produce a stone that appears dark at the centre and in some extreme cases the ring settings may show through the top of the diamond as shadows.

Several different theories on the "ideal" proportions of a diamond have been and continue to be advocated by various owners of patents on machines to view how well a diamond is cut. These advocate a shift away from grading cut by the use of various angles and proportions towards measuring the performance of a cut stone. A number of specially modified viewers and machines have been developed towards this end. Hearts and Arrows viewers test for the "hearts and arrows" characteristic pattern observable in stones exhibiting high symmetry and particular cut angles. Closely related to Hearts and Arrows viewers are the FireScope (also known as the SymmetriScope), the Ideal-Scope, and the ASET. The Ideal-Scope and the ASET test for light leakage, light return, and proportions. The ASET (and computer simulations of the ASET) are used to test for AGS cut grade. The GemEx BrillianceScope is a machine that measures a diamond's white light (brilliance), colour light (fire) and scintillation (sparkle). These viewers and machines often help sellers demonstrate the light performance results of the diamond in addition to the traditional 4 Cs. Detractors see these machines as marketing tools rather than as scientific tools.

The GIA has developed a set of criteria for grading the cut of round brilliant stones that is now the standard in the diamond industry and is called Facetware.

Process

The process of shaping a rough diamond into a polished gemstone is both an art and a science. The choice of cut is often decided by the original shape of the rough stone, location of the inclusions and flaws to be eliminated, the preservation of the weight, popularity of certain shapes amongst consumers and many other considerations. The round brilliant cut is preferred when the crystal is an octahedron, as often two stones may be cut from one such crystal. Oddly shaped crystals such as macles are more likely to be cut in a *fancy cut*—that is, a cut other than the round brilliant—which the particular crystal shape lends itself to.

Even with modern techniques, the cutting and polishing of a diamond crystal always results in a dramatic loss of weight; rarely is it less than 50 per cent. Sometimes the cutters compromise and accept lesser proportions and symmetry in order to avoid inclusions or to preserve the carat rating. Since the per carat price of diamond shifts around key milestones (such as 1.00 carat), many one-carat diamonds are the result of compromising "Cut" for "Carat." Some jewellery experts advise consumers to buy a 0.99 carat diamond for its better price or buy a 1.10 carat diamond for its better cut, avoiding a 1.00 carat diamond which is more likely to be a poorly cut stone.

Light Performance

In the gem trade the term light performance is used to describe how well a polished diamond will return light to the viewer. There are three light properties which are described in relation to light performance; brilliance, fire, and scintillation. Brilliance refers to the white light reflections from the external and internal facet surfaces. Fire refers to the spectral colours which are produced as a result of the diamond dispersing the white light. Scintillation refers to the small flashes of light that are seen when the diamond, light source or the viewer is moved. A diamond that is cut and polished to produce a high level of these qualities is said to be high in *light performance*.

The setting diamonds are placed in also affect the performance of light through a diamond. The 3 most commonly used settings are: Prong, Bezel, and Channel. Prong settings are the most popular setting for diamond jewellery. The prong setting consists of four or six 'claws' that cradle the diamond, allowing the maximum amount of light to enter from all angles, allowing the diamonds to appear larger and more brilliant. In *bezel settings* the diamond or gemstone is completely surrounded by a rim of metal, which can be molded into any shape to accommodate the stone. Used to set earrings, necklaces, bracelets, and rings, bezel settings can have open or closed backs, and generally can be molded to allow a lot of light to pass through. Channel settings set the stones right next to each other with no metal separating them. This setting is mostly used in wedding and anniversary bands. The outer ridge is then worked over the edges of the stones to create a smooth exterior surface. This also protects the girdle area of the stone.

FLUORESCENCE

About a third of all diamonds will glow under ultraviolet light, usually a blue colour which may be noticeable under a black light or strong sunlight. According to the GIA, who reviewed a random sample of 26,010 natural diamonds, 65 per cent of the diamonds in the sample had no fluorescence. Of the 35 per cent that did have fluorescence, 97 per cent had blue fluorescence of which 38 per cent had faint blue fluorescence and 62 per cent had fluorescence that ranged from medium to very strong blue.

Other colours diamonds have fluoresce of green, yellow, and red but are very rare and are sometimes a combination of the colours such as blue-green or orange. Some diamonds with "very strong" fluorescence can have a "milky" or "oily" look to them, but they are also very rare and are termed "overblues." Their study concluded that with the exception of "overblues" and yellow fluorescent diamonds, fluorescence had little effect on transparency and that the strong and very strong blue fluorescent diamonds on average had better colour appearance than non-fluorescent stones. Since blue is a complementary colour to yellow and can appear to cancel it out, strong blue fluorescence had especially better colour appearance with lower colour graded diamonds that have a slight yellowish tint such as "I" colour or "J" colour but had little effect on the more colourless "D" through "F" colour grades.

CLEANING

Cleanliness heavily affects a diamond's beauty. A clean diamond is more brilliant and fiery than the same diamond when it is "dirty". Dirt or grease on the top of a diamond reduces its lustre. Water, dirt, or grease on the bottom of a diamond interferes with the diamond's brilliance and fire. Even a thin film absorbs some light that could have been reflected to the person looking at the diamond. Coloured dye or smudges can affect the perceived colour of a diamond. Historically, some jewellers' stones were misgraded because of smudges on the girdle, or dye on the culet. Current practice is to clean a diamond thoroughly before grading its colour.

Maintaining a clean diamond can sometimes be difficult as jewellery settings can obstruct cleaning efforts and oils, grease, and other hydrophobic materials adhere well to a diamond's surface. Many jewellers use steam cleaners. Some jewellers provide their customers with ammonia-based cleaning kits; ultrasonic cleaners are also popular.

History of Diamonds

Diamonds are thought to have been first recognised and mined in India, where significant alluvial deposits of the stone could then be found. The earliest written reference can be found in the Buddhist text, the Anguttara Nikaya another sanskrit text, the Arthashastra, which was completed around 296 BCE and describes diamond's hardness, lustre, and dispersion. Diamonds quickly became associated with divinity, being used to decorate religious icons, and were believed to bring good fortune to those who carried them. Ownership was restricted among various castes by colour, with only kings being allowed to own all colours of diamond.

In February 2005, a joint Chinese-US team of archaeologists reported the discovery of four corundum-rich stone ceremonial burial axes originating from China's Liangzhu and Sanxingcun cultures (4000 BCE–2500 BCE) which, because of the axes' spicular surfaces, the scientists believe were polished using diamond powder. Although there are diamond deposits now known to exist close to the burial sites, no direct evidence of coeval diamond mining has been found: the researchers came to this conclusion by polishing corundum using various lapidary abrasives and modern techniques then comparing the results using an atomic force microscope. At that scale, the surface of the modern diamond-polished corundum closely resembled that of the axes; however, the polishes of the latter were superior.

Diamonds were traded to both the east and west of India and were recognised by various cultures for their gemological or industrial uses. In his work *Naturalis Historia*, the Roman writer Pliny the Elder noted diamond's ornamental uses, as well as its usefulness to engravers because of its hardness. In China, diamonds seem to have been used primarily as diamond tools for engraving jade and drilling holes in beads. Archaeological evidence from Yemen suggests that diamonds were used as drill tips as early as the 4th century BCE. In Europe, however, diamonds disappeared for almost 1,000 years following the rise of Christianity because of two effects: early Christians rejected diamonds because of their earlier use in amulets, and Arabic traders restricted the flow of trade between Europe and India.

Until the late Middle Ages, diamonds were most prized in their natural octahedral state, perhaps with the crystal surfaces polished to increase lustre and remove foreign material. Around 1300, the flow of diamonds into Europe increased via Venice's trade network, with most flowing through the low country ports of Bruges, Antwerp, and Amsterdam. During this time, the taboo against cutting diamonds into gem shapes, which was established over 1,000 years earlier in the traditions of India, ended allowing the development of diamond cutting technology to begin in earnest. By 1375, a guild of diamond polishers had been established at Nuremberg. Over the following centuries, various diamond cuts were introduced which increasingly demonstrated the fire and brilliance that makes diamonds treasured today: the *table cut*, the *briolette* (around 1476), the *rose cut* (mid-16th century), and by the mid-17th century, the *Mazarin*, the first brilliant cut diamond design. In 1919, Marcel Tolkowsky developed an *ideal* round brilliant cut design that has set the standard for comparison of modern gems; however, diamond cuts have continued to be refined.

The rise in popularity of diamonds as gems seems to have paralleled increasing availability through European history. In the 13th century, King Louis IX of France established a law that only the king could own diamonds. However, within a century diamonds were popular gems among the moneyed aristocratic and merchant classes, and by 1477 had begun to be used rarely in wedding rings. Diamond wedding rings didn't gain widespread social significance until the De Beers company started marketing the idea through cinema beginning in the 1940s. A number of large diamonds have become historically significant objects, as their inclusion in various sets of crown jewels and the purchase, sale, and sometimes theft of notable diamonds, have sometimes become politicised.

Record-holding Diamonds

The Cullinan Diamond, part of the British crown jewels, was the largest gem-quality rough diamond ever found (1905), at 3,106.75 carats. One of the diamonds cut from it, Cullinan I or the Great Star of Africa, was formerly the largest gem-quality cut diamond at 530.2 carats, but now that title has been taken by the Golden Jubilee (1985), a 545.67 carat, yellow-brown diamond. The largest flawless and colourless (grade D) diamond is the Centenary Diamond which weighs 273.85 carats. The Millennium Star is the second largest (1990) at 203.04 carats.

Travis Metcalfe, an astronomer at the Harvard-Smithsonian Centre for Astrophysics, believes that the galaxy's largest diamond is the core of the white dwarf star BPM 37093. Observations indicate that the core is a diamond crystal 4000 km in diameter.

Diamond Industry

The diamond industry can be broadly separated into two basically distinct categories: one dealing with gem-grade diamonds and another for industrial-grade diamonds. While a large trade in both types of diamonds exists, the two markets act in dramatically different ways.

Gem Diamond Industry

A large trade in gem-grade diamonds exists. Unlike precious metals such as gold or platinum, gem diamonds do not trade as a commodity: there is a substantial mark-up in the sale of diamonds, and there is not a very active market for resale of diamonds. One hallmark of the trade in gem-quality diamonds is its remarkable concentration: wholesale trade and diamond cutting is limited to a few locations (most importantly New York, Antwerp, London, Tel Aviv, Amsterdam and Surat), and a single company—De Beers—controls a significant proportion of the trade in diamonds. They are based in Johannesburg, South Africa and London, England.

The production and distribution of diamonds is largely consolidated in the hands of a few key players, and concentrated in traditional diamond trading centres. The most important being Antwerp, where 80 per cent of all rough diamonds, 50 per cent of all cut diamonds and more than 50 per cent of all rough, cut and industrial diamonds combined are handled. This makes Antwerp the de facto 'world diamond capital'.

The De Beers company, as the world's largest diamond miner holds a clearly dominant position in the industry, and has done so since soon after its founding in 1888 by the British imperialist Cecil Rhodes. De Beers owns or controls a significant portion of the world's rough diamond production facilities (mines) and distribution channels for gem-quality diamonds. The company and its subsidiaries own mines that produce some 40 per cent of annual world diamond production. At one time it was thought over 80 per cent of the world's rough diamonds passed through the Diamond Trading Company (DTC, a subsidiary of De Beers) in London, but presently the figure is estimated at less than 50 per cent. De Beers used its monopoly position to establish strict price controls, and market diamonds directly to consumers in world markets.

The De Beers diamond advertising campaign is acknowledged as one of the most successful and innovative campaigns in history. N.W. Ayer & Son, the advertising firm retained by De Beers in the mid-20th century, succeeded in reviving the American diamond market and opened up new markets, even in countries where no diamond tradition had existed before. N.W. Ayer's multifaceted marketing campaign included product placement, advertising the diamond itself rather than the De Beers brand, and building associations with celebrities and royalty. This coordinated campaign has lasted decades and continues today; it is perhaps best captured by the slogan "a diamond is forever". The De Beers account is now handled by the firm JWT, formerly known as J. Walter Thompson.

Further down the supply chain, members of The World Federation of Diamond Bourses (WFDB) act as a medium for wholesale diamond exchange, trading both polished and rough diamonds. The WFDB consists of independent diamond bourses in major cutting centres such as Tel Aviv, Antwerp, Johannesburg and other cities across the USA, Europe and Asia.

In 2000, the WFDB and The International Diamond Manufacturers Association established the World Diamond Council to prevent the trading of diamonds used to fund war and inhumane acts.

WFDB's additional activities also include sponsoring the World Diamond Congress every two years, as well as the establishment of the *International Diamond Council* (IDC) to oversee diamond grading.

Industrial Diamond Industry

The market for industrial-grade diamonds operates much differently from its gem-grade counterpart. Industrial diamonds are valued mostly for their hardness and heat conductivity, making many of the gemological characteristics of diamond, including clarity and colour, mostly irrelevant. This helps explain why 80 per cent of mined diamonds (equal to about 100 million carats or 20,000 kg annually), unsuitable for use as gemstones and known as *bort*, are destined for industrial use. In addition to mined diamonds, synthetic diamonds found industrial applications almost immediately after their invention in the 1950s; another 3 billion carats (600 metric tons) of synthetic diamond is produced annually for industrial use—nearly 25 times the mass of natural diamonds mined over the same period.

The dominant industrial use of diamond is in cutting, drilling, grinding, and polishing. Most uses of diamonds in these technologies do not require large diamonds; in fact, most diamonds that are gem-quality except for their small size, can find an industrial use. Diamonds are embedded in drill tips or saw blades, or ground into a powder for use in grinding and polishing applications. Specialised applications include use in laboratories as containment for high pressure experiments, high-performance bearings, and limited use in specialised windows.

With the continuing advances being made in the production of synthetic diamonds, future applications are beginning to become feasible. Garnering much excitement is the possible use of diamond as a semi-conductor suitable to build microchips from, or the use of diamond as a heat sink in electronics. Significant research efforts in Japan, Europe, and the United States are under way to capitalise on the potential offered by diamond's unique material properties, combined with increased quality and quantity of supply starting to become available from synthetic diamond manufacturers.

Diamond Supply Chain

The diamond supply chain is controlled by a limited number of powerful businesses, and is also highly concentrated in a small number of locations around the world. In fact, the amount of power which De Beers has consolidated historically prevented it from direct trade with the United States, as its trade practices led to an indictment for violating anti-trust regulations (the case was settled in 2004). The concentration of power only loosens at the retail level, where diamonds are sold by a limited number of distributors, known as sightholders, to jewellers around the world.

Sources: Historically diamonds were known to be found only in alluvial deposits in southern India; India led the world in diamond production from the time of their discovery in approximately the 9th century BCE to the mid-18th century AD, but the commercial potential of these sources has been exhausted. The first non-Indian diamond source was found in Brazil in 1725. While no commercial diamond production exists in the US, Arkansas and Colourado are the only states to have a verifiable source of diamonds. Today, most commercially viable diamond deposits are in Africa, notably in South Africa, Namibia, Botswana, the Democratic Republic of Congo, Angola, Tanzania and Sierra Leone. There are also commercial deposits being actively mined in the Northwest Territories of Canada, Siberia (mostly in Yakutia territory, for example Mir pipe and Udachnaya pipe), Brazil, and in Northern and Western Australia. Diamond prospectors continue to search the globe for diamond-bearing kimberlite and lamproite pipes.

Blood Diamonds

In some of the more politically unstable central African and west African countries, revolutionary groups have taken control of diamond mines, using proceeds from diamond sales to finance their operations. Diamonds sold through this process are known as *conflict diamonds* or *blood diamonds*. In response to public concerns that their diamond purchases were contributing to war and human rights abuses in central Africa and west Africa, the United Nations, the diamond industry and diamond-trading nations introduced the Kimberley Process in 2002, which is aimed at ensuring that conflict diamonds do not become intermixed with the diamonds not controlled by such rebel groups. The Kimberley Process provides documentation and certification of diamond exports from producing countries to ensure that the proceeds of sale are not being used to fund criminal or revolutionary activities. Although the Kimberley Process has been moderately successful in limiting the number of conflict diamonds entering the market, conflict diamonds smuggled to market continue to persist to some degree (approx. 2-3 per cent of diamonds traded today are possible conflict diamonds). According to the 2006 book *The Heartless Stone*, two major flaws still hinder the effectiveness of the Kimberley Process: the relative ease of smuggling diamonds across African borders and giving phony histories, and the violent nature of diamond mining in nations which are not in a technical state of war and whose diamonds are therefore considered "clean."

The Canadian Government has setup a body known as Canadian Diamond Code of Conduct: to help authenticate Canadian Diamonds. This is a very stringent tracking system of diamonds and helps protect the 'conflict free' label of Canadian diamonds.

Currently, gem production totals nearly 30 million carats (6,000 kg) of cut and polished stones annually, and over 100 million carats (20,000 kg) of mined diamonds are sold for industrial use each year, as are about 100,000 kg of synthesised diamond. In 2003, this constituted total production of nearly US$9 billion in value.

Mining

Only a very small fraction of the diamond ore consists of actual diamonds. The ore is crushed, during which care has to be taken in order to prevent larger diamonds from being destroyed in this process and subsequently the particles are sorted by density. Nowadays, the diamonds are located in the diamond-rich density fraction with the help of X-ray fluorescence, after which the final sorting steps are done by hand. Before the use of X-rays became commonplace, the separation was done with grease belts; diamonds have a stronger tendency to stick to grease than the other minerals in the ore.

Distribution

The Diamond Trading Company, or DTC, is a subsidiary of De Beers and markets rough diamonds produced both by De Beers mines and other mines from which it purchases rough diamond production. DTC performs sophisticated sorting of rough diamonds into over 16,000 categories, and then sells bulk lots of rough diamonds to a limited number of sightholders a few times a year.

Once purchased by sightholders, diamonds are cut and polished in preparation for sale as gemstones. The cutting and polishing of rough diamonds is a specialised skill that is concentrated in a limited number of locations worldwide. Traditional diamond cutting centres are Antwerp, Amsterdam, Johannesburg, New York, and Tel Aviv. Recently, diamond cutting centres have been established in China, India, and Thailand. Cutting centres with lower costs of labour, notably Surat in Gujarat, India, handle a larger number of smaller carat diamonds, while smaller quantities of larger or more valuable diamonds are more likely to be handled in Europe or North America. Demonstrating this, India produces 90 per cent of all cut and polished diamonds by number, but only 55 per cent by value. The recent expansion of this industry in India, employing low cost labour, has allowed smaller diamonds to be prepared as gems than was previously economically feasible.

Diamonds which have been prepared as gemstones are sold on diamond exchanges called *bourses*. There are 24 registered diamond bourses. This is the final tightly controlled step in the diamond supply chain; wholesalers and even retailers are able to buy relatively small lots of diamonds at the bourses, after which they are prepared for final sale to the consumer. Diamonds can be sold already set in jewellery, or as is increasingly popular, sold unset ("loose"). According to the Rio Tinto Group, in 2002 the diamonds produced and released to the market were valued at US$9 billion as rough diamonds, US$14 billion after being cut and polished, US$28 billion in wholesale diamond jewellery, and retail sales of US$57 billion.

Synthetics, Simulants, and Enhancements

It is important to distinguish that a synthetic diamond is a true diamond created by a technological process, whereas a diamond simulant is defined as a non-diamond material that is used to simulate the appearance of a true diamond.

The gemological and industrial uses of diamond have created a large demand for rough stones. A portion of this demand is now being met by synthetic diamonds, artificially-made diamonds which have the properties of natural diamonds. This process has historically produced industrial-grade diamonds, but synthetic diamond producers have recently begun to produce diamonds with high enough quality to penetrate the gem diamond market. Diamonds have been manufactured synthetically for over fifty years.

A diamond's gem quality, which is not as dependent on material properties as industrial applications, has invited both imitation and the invention of procedures to enhance the gemological properties of natural diamonds. Materials which have similar gemological characteristics to diamond but are not mined or synthetic diamond are known as *diamond simulants*. The most familiar diamond simulant to most consumers is cubic zirconia (commonly abbreviated as CZ); recently moissanite has also gained cachet but has often been mischaracterised as a diamond simulant although it is sold and retailed as a replacement for diamond. Both CZ and moissanite are synthetically produced. However CZ is a diamond simulant. Diamond enhancements are specific treatments, performed on natural diamonds (usually those already cut and polished into a gem), which are designed to better the gemological characteristics of the stone in one or more ways. These include laser drilling to remove inclusions, application of sealants to fill cracks, treatments to improve a white diamond's colour grade, and treatments to give fancy colour to a white diamond.

Currently, trained gemologists with appropriate equipment are able to distinguish natural diamonds from all synthetic and simulant diamonds, and identify all enhanced natural diamonds. The established natural diamond industry has a vested interest in maintaining the distinction between natural diamonds and other diamonds, and has made significant investments towards that end. However, as manufacturing technology improves, synthetic diamonds may become indistinguishable from natural diamonds, and new techniques for creating and treating simulants (such as coating them with a very thin diamond-like layer of carbon) are making it increasingly difficult to distinguish simulants from real diamonds.

Symbolism

Historically, it has been claimed that diamonds possess several supernatural powers:

- A diamond gives victory to him who carries it bound on his left arm, no matter the number of enemies.
- Panics, pestilences, enchantments, all fly before it; hence, it is good for sleepwalkers and the insane.
- It deprives lodestone and magnets of their virtue (i.e., ability to attract iron).
- Arabic diamonds are said to attract iron greater than a magnet.
- A diamond's hardiness can only be broken by smearing it with fresh goat's blood.

Because of their extraordinary physical properties, diamonds have been used symbolically since near the time of their first discovery. Perhaps the earliest symbolic use of diamonds was as the eyes of Hindu devotional statues. In Hinduism, Indra uses Vajrayudham or the thunderbolt as his primary weapon. *Vajra* is the word for diamond and *ayudham* means weapon in Sanskrit. The diamonds themselves were thought to be endowments from the gods and were therefore cherished. The point at which diamonds began to be associated with divinity is not known, but early texts indicate that it was recognised in India since at least 400 BCE. It is said the Greeks believed diamonds were tears of the gods; the Romans believed they were splinters of fallen stars. Many long dead cultures have sought to explain diamond's superlative properties through divine or mystical affiliations.

In Tibetan Buddhism, also known as Vajrayana (Diamond Vehicle), diamonds are an important symbol, and the Diamond Sutra is one of the most popular texts.

In Western culture, diamonds are the traditional emblem of fearlessness and virtue, but have also often associated with power, wealth, crime and misfortune. Today, diamonds are used to symbolise eternity and love, being often seen adorning engagement rings and sometimes wedding rings as well. The popularity of this modern tradition can be traced directly to the marketing campaigns of De Beers, starting in 1938. Prior to the De Beers marketing campaign, engagement rings had no one particular stone associated with them. The first diamond engagement ring can be traced to the marriage of Maximilian I (then Archduke of Austria) to Mary of Burgundy in 1477. Other early examples of betrothal jewels incorporating diamonds include the *Bridal Crown of Blanche* (ca. 1370–80) and the *Heftlein* brooch of Vienna (ca. 1430–40), a pictorial piece depicting a wedding couple. Inaccessibility of diamonds to the vast majority of the population limited the popularity of diamonds as betrothal jewels during this period.

The diamond is the birthstone for people born in the month of April, and is also used as the symbol of a sixty-year anniversary, such as a Diamond Jubilee. In a system of heraldry by gemstones occasionally used in the past for the arms of nobles, diamond was used to represent the colour sable, or black.

Diamonds are a common focus of fiction. Notable pieces of fiction include Ian Fleming's *Diamonds Are Forever* (1956), Arthur C. Clarke's *2061: Odyssey Three* (1988), F. Scott Fitzgerald's "The Diamond as Big As the Ritz" (1922), and Neal Stephenson's *The Diamond Age* (1995). In addition, diamonds are the subject of various myths and legends.

EMERALD

Emerald ($Be_3Al_2(SiO_3)_6$) is a variety of the mineral **beryl,** coloured green by trace amounts of chromium and sometimes vanadium. It is highly prized as a gemstone and by weight is the most valuable gemstone in the world, although it is often made less so by inclusions, which all emeralds have to some degree. Beryl has a hardness of 7.5 - 8 on the 10 point Mohs scale of hardness. However, the effective hardness of an emerald is often reduced by fractures and inclusions. Most emeralds are highly included, so the toughness (resistance to breakage) is classified as generally poor.

Emeralds come in many shades of green and bluish green. There is a wide spectrum of clarity, dependent on the inclusions and fractures in the crystal. Clear stones with dark yet vibrant colour command the highest prices.

Most emeralds are oiled as part of the post-lapidary process, in order to improve their clarity. Cedar oil is often used, having a similar refractive index, and it is a generally accepted practice. The Federal Trade Commission requires the disclosure of this type of treatment when a treated emerald is sold. The amount of oil entering an emerald microfissure is roughly equivalent to the size of a period (full stop) in print. The use of green tinted oil is generally not considered acceptable by the gem trade.

Emeralds in antiquity were mined by the Egyptians and in Austria, as well as Swat in northern Pakistan.

A rare type of emerald known as a trapiche emerald is occasionally found in the mines of Colombia. A trapiche emerald exhibits a "star" pattern; it has raylike spokes of dark carbon impurities that give the emerald a six-pointed radial pattern. It is named for the *trapiche*, a grinding wheel used to process sugarcane in the region. Colombian emeralds are generally the most prized due to their transparency and fire. Some of the most rare emeralds come from three main emerald mining areas in Colombia: Muzo, Coscuez, and Chivor. Fine emeralds are also found in other countries, such as Zambia, Brazil, Zimbabwe, Madagascar, Pakistan, India, Afghanistan and Russia.

The value of an emerald depends on cut, colour, clarity, and carat. The characteristics of Colombian emeralds set the highest standards of quality.

Emerald is a rare and valuable gemstone and, as such, it has provided the incentive for developing synthetic emeralds. Both hydrothermal and *flux-growth* synthetics have been produced, and a method has been developed for producing an emerald overgrowth on colourless beryl. The first commercially successful emerald synthesis process was that of Carroll C. Chatham. Because Chatham's emeralds do not have any water and contain traces of vanadate, molybdenum and vanadium, a lithium vanadate flux process is probably involved. The other large producer of flux emeralds is Pierre Gilson Sr., which has been on the market since 1964. Gilson's emeralds are usually grown on natural colourless beryl seeds which become coated on both sides. Growth occurs at the rate of 1 mm per month and a typical seven-month growth run produces emerald crystals of 7 mm of thickness (Nassau, K. Gems Made By Man, 1980).

Hydrothermal synthetic emeralds have been attributed to IG-Farben, Nacken, Tairus and others, but the first satisfactory commercial product was that of Johann Lechleitner of Inbruck, Austria, which appeared on the market in the 1960s. These stones were initially sold under the names "Emerita" and "Symeralds", and they were grown as a thin layer of emerald on top of natural colourless beryl stones.

Although not much is known about the original process, it is assumed that Leichleitner emeralds were grown on acid conditions. Later, from 1965 to 1970, the Linde Division of Union Carbide produced completely synthetic emeralds by hydrothermal synthesis. According to their patents (US3,567,642 and US3,567,643), acidic conditions are essential to prevent the chromium (which is used as the colourant) from precipitating.

Also, it is important that the silicon containing nutrient be kept away from the other ingredients in order to prevent nucleation and confine growth to the seed crystals. Growth occurs by a diffusion-reaction process, assisted by convection. Typical growth conditions include pressures of 700-1400 bars at temperatures of 500 to 600°C with a temperature gradient of 10 to 25°C. Growth rates as fast as 1/3 mm per day can be attained.

Luminescence in ultraviolet light is considered a supplementary test when making a natural vs. synthetic determination, as many, but not all, natural emeralds are inert to ultraviolet light. Many synthetics are also UV inert.

Synthetic emeralds are often referred to as "created", as their chemical and gemological composition is the same as their natural counterparts. The Federal Trade Commission (FTC) has very strict regulations as to what can and what cannot be called "synthetic" stone. The FTC says: It is unfair or deceptive to use the word "laboratory-grown," "laboratory-created," "[manufacturer name]-created," or "synthetic" with the name of any natural stone to describe any industry product unless such industry product has essentially the same optical, physical, and chemical properties as the stone named."

Today there are only four commercial laboratories growing synthetic emeralds, all of them are in Russia with one, Tairus, contributing to about 90 per cent of world's production.

Wispy veil-like inclusions are common in flux-grown synthetic emeralds.

OPAL

The mineraloid opal is amorphous $SiO_2 \cdot nH_2O$; hydrated silicon dioxide, the water content sometimes being as high as 20 per cent but is usually between three and ten per cent. Opal ranges from colourless through white, milky blue, gray, red, yellow, green, brown and black. Common opal is truly amorphous, but precious opal does have a structural element. The word opal comes from the Sanskrit upala, the Greek opallios, and the Latin opalus, meaning "precious stone." Opals are also Australia's National gemstone.

Opal is a mineraloid gel which is deposited at relatively low temperature and may occur in the fissures of almost any kind of rock, being most commonly found with limonite, sandstone, rhyolite, and basalt.

Opal is one of the mineraloids that can form or replace fossils. The resulting fossils, though not of any extra scientific interest, appeal to collectors.

PRECIOUS OPAL

Precious opal shows a variable interplay of internal colours and does have an internal structure. At the micro scale precious opal is composed of silica spheres some 150 to 300 nm in diameter in a hexagonal or cubic closed-packed lattice. These ordered silica spheres produce the internal colours by causing the interference and diffraction of light passing through the microstructure of opal (Klein and Hurlbut, 1985, p. 444). It is the regularity of the sizes of the spheres, and of the packing of these spheres that determines the quality of precious opal. Where the distance between the regularly packed planes of spheres is approximately half the wavelength of a component of visible light, the light of that wavelength may be subject to diffraction from the grating created by the stacked planes.

The spacing between the planes and the orientation of planes with respect to the incident light determines the colours observed. The process can be described by Bragg's Law of diffraction. Visible light of diffracted wavelengths cannot pass through large thicknesses of the opal. This is the basis of the optical band-gap in a photonic crystal, of which opal is the best known natural example.

Precious opal consists of spheres of silica of fairly regular size, packed into close-packed planes which are stacked together with characteristic dimensions of several hundred nm.

In addition, microfractures may be filled with secondary silica and form thin lamellae inside the opal during solidification. The term *opalescence* is commonly and erroneously used to describe this unique and beautiful phenomenon, which is correctly termed *play of colour*. Contrarily, *opalescence* is correctly applied to the milky, turbid appearance of common or *potch* opal. Potch does not show a play of colour.

The veins of opal displaying the play of colour are often quite thin, and this has given rise to unusual methods of preparing the stone as a gem. An opal *doublet* is a thin layer of colourful material, backed by a black mineral, such as ironstone, basalt or obsidian.

The darker backing emphasises the play of colour, and results in a more attractive display than a lighter potch. Given the texture of opals, they can be quite difficult to polish to a reasonable lustre. The *triplet* cut backs the coloured material with a dark backing, and then has a cap of clear quartz (rock crystal) on top, which takes a high polish, and acts as a protective layer for the comparatively delicate opal.

Common Opal

Besides the gemstone varieties that show a play of colour, there are other kinds of common opal such as the milk opal, milky bluish to greenish (which can sometimes be of gemstone quality); resin opal, honey-yellow with a resinous lustre; wood opal, caused by the replacement of the organic material in wood with opal; menilite brown or grey; hyalite, a colourless glass-clear opal sometimes called Muller's Glass; geyserite, (siliceous sinter) deposited around hot springs or geysers; and diatomite or diatomaceous earth, the accumulations of diatom shells or tests.

Other Varieties of Opal

Fire opal, or Girasol, is a translucent to semi-opaque stone that is generally yellow to bright orange and sometimes nearly red and displays pleochroism at certain angles.

Peruvian opal (also called blue opal) is a semi-opaque to opaque blue-green stone found in Peru which is often cut to include the matrix in the more opaque stones. It does not display pleochroism.

Sources of Opal

Australia produces around 97 per cent of the world's opal. 90 per cent is called 'light opal' or white and crystal opal. White makes up 60 per cent and all the opal fields produce white opal; Crystal opal or pure hydrated silica makes up 30 per cent; 8 per cent is black and only 2 per cent is boulder opal.

The town of Coober Pedy in South Australia is a major source of opal. Another Australian town, Lightning Ridge in New South Wales, is the main source of black opal, opal containing a predominantly dark background (dark-gray to blue-black displaying the *play of colour*). *Boulder opal* consists of concretions and fracture fillings in a dark siliceous ironstone matrix. It is found sporadically in western Queensland, from Kynuna in the north, to Yowah and Koroit in the south.

Fire Opal is found mostly in Mexico and Mesoamerica. In South America, a city called Pedro II, located in Brazil, produces opal that was discovered in 1930. In Honduras there was also some fine *black opal* mined from volcanic ash deposits. This opal is known for its stability.

The Virgin Valley Opal Fields of Humboldt county in northern Nevada produce a wide variety black, crystal, white, and fire opal. Most precious opals are wood replacements. Many specimens have a high water content, and as a result, have a greater tendency to desiccate and crack than most precious opal. Discovered in 1904 the mines are still producing gem materials in large amounts to hundreds of seasonal visitors. Three Fee Dig Mines provide the general public an opportunity to dig the gems themselves. The largest black opal in the Smithsonian Museum, possibly worth in excess of $1 million, comes from the Royal Peacock Opal Mine in the Virgin Valley.

Another source of *white base opal* in the United States is Spencer, Idaho. A high percentage of the opal found there occurs in thin layers. As a result, most of the production goes into the making of doublets and triplets.

Synthetic Opal

As well as occurring naturally, opals of all varieties have been synthesised experimentally and commercially. The discovery of the *ordered sphere* structure of precious opal led to its synthesis by Pierre Gilson in 1974 (Klein and Hurlbut, 1985, p.528). The resulting material is distinguishable from natural opal by its regularity; under magnification, the patches of colour are seen to be arranged in a "lizard skin" or "chicken wire" pattern. Synthetics are further distinguished from naturals by the former's lack of fluorescence under UV light.

Synthetics are also generally lower in density and are often highly porous; some may even stick to the tongue.

Two notable producers of synthetic opal are the companies Kyocera and Inamori of Japan. Most so-called synthetics, however, are more correctly termed *imitations,* as they contain substances not found in natural opal (e.g., plastic stabilisers). The imitation opals seen in vintage jewellery are often "Slocum Stone" consisting of laminated glass with bits of foil interspersed.

Local Atomic Structure of Opals

The lattice of spheres of opal that cause the interference with light are several hundred times larger than the fundamental structure of crystalline silica. As a mineraloid, there is no unit cell that describes the structure of opal. Nevertheless, opals can be roughly divided into those that have shown no signs of crystalline order, i.e., amorphous opal, and those that show signs of the beginning of crystalline order, commonly termed cryptocrystalline or microcrystalline opal (Graetsch, 1994).

Dehydration experiments and infrared spectroscopy have shown that most of the H_2O in the formula of $SiO_2.nH_2O$ of opals is present in the familiar form of clusters of molecular water. Isolated water molecules, and silanols, structures such as Si-O-H, generally form a lesser proportion of the total and can reside near the surface or in defects inside the opal.

The structure of low-pressure polymorphs of anhydrous silica consist of frameworks of fully-corner bonded tetrahedra of SiO_4. The higher temperature polymorphs of silica cristobalite and tridymite are frequently the first to crystallise from amorphous anhydrous silica, and the local structures of microcrystalline opals also appear to be closer to that of cristobalite and tridymite than to quartz. The structures of tridymite and cristobalite are closely related and can be described as hexagonal and cubic close-packed layers. It is therefore possible to have intermediate structures in which the layers are not regularly stacked.

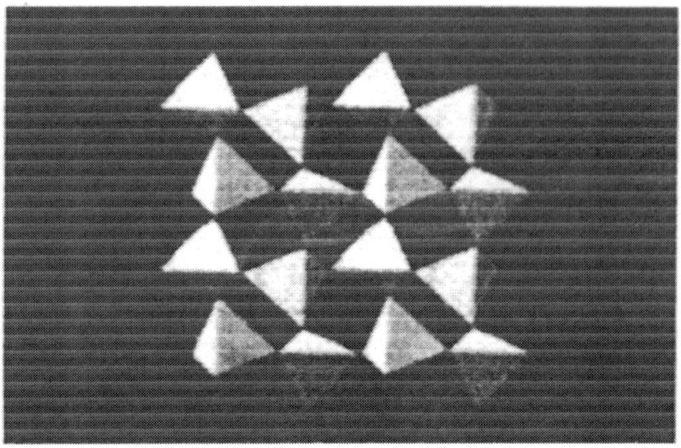

The crystal structure of crystalline a-cristobalite. Locally, the structures of some opals, opal-C, are similar to this

Microcrystalline Opal

Opal-CT has been interpreted as consisting of clusters of stacking of cristobalite and tridymite over very short length scales. The spheres of opal in opal-CT are themselves made up of tiny microcrystalline blades of cristobalite and tridymite. Opal-CT has occasionally been further subdivided in the literature. Water content may be as high as 10 wt per cent.

Opal-C is interpreted as consisting of localised order of a-cristobalite with a lot of stacking disorder. Typical water content is about 1.5wt per cent.

Non-crystalline Opal

Two broad categories of non-crystalline opals, sometimes just referred to as opal-A, have been proposed.

Opal-AG: Aggregated spheres of silica, with water filling the space in between. Precious opal and potch opal are generally varieties of this, the difference being in the regularity of the sizes of the spheres and their packing.

Opal-AN: Water-containing amorphous silica-glass. Hyalite is a synonym.

Non-crystalline silica in siliceous sediments is reported to gradually transform to opal-CT and then opal-C as a result of diagenesis, due to the increasing overburden pressure in sedimentary rocks, as some of the stacking disorder is removed (Cady *et al.*, 1996).

Ruby

Ruby is a red gemstone that varies from a light pink to a blood red, a variety of the mineral corundum (aluminium oxide). The colour is caused mainly by chromium. Its name comes from rubber, Latin for red. Natural rubies are exceptionally rare, but synthetic rubies (sometimes called created ruby) can be manufactured fairly cheaply. Other varieties of gem-quality corundum are called sapphires. It is considered one of the four precious gems, together with the sapphire, the emerald and the diamond.

Rubies are mined in Africa, Asia, Australia, Greenland, Madagascar and North Carolina. They are most often found in Myanmar (Burma), Sri Lanka, Kenya, Madagascar, and Cambodia, but they have also been found in the US states of Montana, North Carolina and South Carolina.

The Mogok Valley in Upper Myanmar has produced some of the finest rubies but, in recent years, very few good rubies have been found there. The unique colour in Myanmar (Burmese) rubies is described as "pigeon's blood". They are known in the trade as "Mogok" rubies.

In central Myanmar the area of Mong Hsu also produces rubies. The latest ruby deposit to be found in Myanmar is situated in Nam Ya. In 2002 rubies were found in the Waseges River area of Kenya. Sometimes spinels are found along with rubies in the same rocks and are mistaken for rubies. However, fine red spinels may approach the average ruby in value.

Rubies have a hardness of 9.0 on the Mohs scale of mineral hardness. Among the natural gems only diamond is harder.

A cut ruby

Ruby gemstones are valued according to size, colour, clarity and cut. All natural rubies have imperfections in them, including colour impurities and inclusions of rutile needles known as "silk". Gemologists use these needle inclusions found in natural rubies to distinguish them from synthetics, simulants, or substitutes. If there is no silk in the stone, that shows that the stone was heated to a temperature of up to 1800°C (3272°F) in an oven to give the ruby a better colour of red. Usually the rough stone is heated before cutting. About 90 per cent of all rubies today are heated, and rubies which are not heated are considered unusual. Some rubies undergo a process of low tube heat, when the stone is heated over charcoal of a temperature of about 1300°C (2372°F) for 20 to 30 minutes. The silk is partially broken and the colour is improved. The fracture filling of rubies is also done intentionally, and it is not always disclosed to gem buyers. Glass-filling voids in rubies, without disclosure, is considered an unethical practice.

General Identification of Gems

Gem identification is basically a process of elimination. Gemstones of similar colour undergo non-destructive optical testing until there is only one possible identity. Any single test is indicative, only. For example, the specific gravity of ruby is 4.00, glass is 3.15-4.20, and cubic zirconia is 5.6-5.9. So, one can easily tell the difference between cubic zirconia and the other two; however, there is overlap between ruby and glass.

And, as with all naturally occurring material(s), no two gems are identical. The geological environment they are created in, influences the overall process so that although the basics can be identified the presence of chemical "impurities" and substitutions along with structural imperfections vary thus creating "individuals".

Identification by Refractive Index

One test to determine the gem's identity is to measure the refraction of light in the gem. Every material has a critical angle, at which point light is reflected back internally. This can be measured and thus used to determine the gem. Typically, this is measured using a refractometer although it is possible to measure it using a microscope.

Identification by Specific Gravity

Specific gravity, also known as relative density, is different depending upon the chemical composition and crystal structure type. Heavy liquids with a known specific gravity are used to test loose gemstones. Specific Gravity is measured by comparing the weight of the gem in air with the weight of the gem suspended in water.

Identification by Spectroscopy

This method uses a similar principle to how a prism works to separate white light into its component colours. A gemological spectroscope is employed to analyse the selective absorption of light in the gem material. Essentially, when light passes from one medium to another it bends. Blue light bends more than red light. Depending upon the gem material will adjust how much this light bends. Colouring agents or chromophores show bands in the spectroscope and indicate what element is responsible for the gems colour. The wavelengths displayed are measured in nanometres.

4

Crystallography

Crystallography is the experimental science of determining the arrangement of atoms in solids. In older usage, it is the scientific study of crystals.

Before the development of X-ray diffraction crystallography, the study of crystals was based on the geometry of the crystals. This involves measuring the angles of crystal faces relative to theoretical reference axes (crystallographic axes), and establishing the symmetry of the crystal in question. The former is carried out using a goniometer. The position in 3D space of each crystal face is plotted on a stereographic net, e.g. Wolff net or Lambert net. In fact, the pole to each face is plotted on the net. Each point is labelled with its Miller index. The final plot allows the symmetry of the crystal to be established.

Crystallographic methods now depend on the analysis of the diffraction patterns that emerge from a sample that is targeted by a beam of some type. The beam is not always electromagnetic radiation, even though X-rays are the most common choice. For some purposes electrons or neutrons are used, which is possible due to the wave properties of the particles. Crystallographers often explicitly state the type of illumination used when referring to a method, as with the terms X-ray diffraction, neutron diffraction and electron diffraction.

These three types of radiation interact with the specimen in different ways. X-rays interact with the spatial distribution of the valence electrons, while electrons are charged particles and therefore feel the total charge distribution of both the atomic nuclei and the surrounding electrons. Neutrons are scattered by the atomic nuclei through the strong nuclear forces, but in addition, the magnetic moment of neutrons is non-zero. They are therefore also scattered by magnetic fields. Because of these different forms of interaction, the three types of radiation are suitable for different crystallographic studies.

THEORY

In several cases, an image of a microscopic object is generated by focusing the rays of the visible spectrum using a lens as in light microscopy. However, because the wavelength of visible light is long compared to atomic bond lengths and atoms themselves, it is necessary to use radiation with shorter wavelengths, such as X-rays. Employing shorter wavelengths implies abandoning microscopy and true imaging, however, because there exists no material from which a lens capable of focusing this type of radiation can be created. (That said, scientists have had some success focusing X-rays with microscopic Fresnel zone plates made from gold). Generally, in diffraction-based imaging, the only wavelengths used are those that are too short to be focused. This difficulty is the reason that crystals must be used.

Because of their highly ordered and repetitive structure, crystals are an ideal material for analysing the structure of solids. To use X-ray diffraction as an example, a single X-ray photon diffracting off one electron cloud not generate a strong enough signal for the equipment to detect. However, many

X-rays diffracting off many electron clouds in approximately the same relative position and orientation throughout the crystal will result in constructive interference and hence a detectable signal.

NOTATION

Miller Index

Miller indices are a notation used to describe lattice planes and directions in a crystal.

In particular, a family of lattice planes is determined by three integers $\ell, m,$ and n, the *Miller indices*. They are written (ℓmn) and denote planes orthogonal to a direction (ℓ, mn) in the basis of the reciprocal lattice vectors. By convention, negative integers are written with a bar, as in $\bar{3}$ for $\bar{3}$. The integers are usually written in lowest terms, i.e. their greatest common divisor should be 1.

The precise meaning of this notation depends upon a choice of lattice vectors for the crystal, as described below. Usually, the three primitive lattice vectors are used. However, for cubic crystal systems, the cubic lattice vectors are used even when they are not primitive (e.g., as in body-centred and face-centred crystals).

There are also several related notations (Ashcroft & Mermin, 1976). $[\ell mn]$, with square instead of round brackets, denotes a direction in the basis of the direct lattice vectors instead of the reciprocal lattice. The notation $[\ell mn]$ denotes all planes that are equivalent to by the symmetry of the crystal. Similarly, the notation denotes all directions that are equivalent to by symmetry.

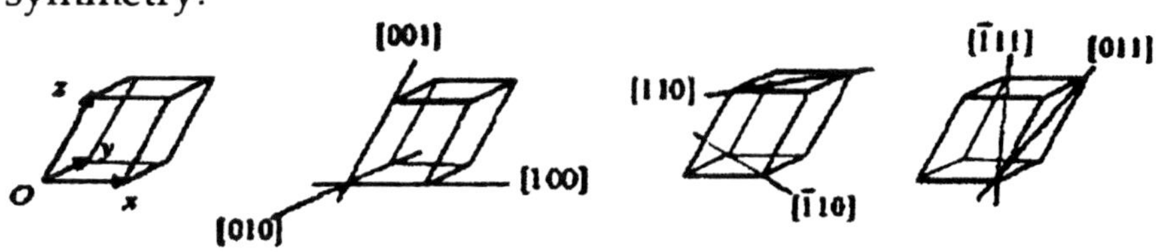

Indices miller direction examples

DEFINITION

There are two equivalent ways to define the meaning of the Miller indices (Ashcroft & Mermin, 1976): via a point in the reciprocal lattice, or as the inverse intercepts along the lattice vectors. Both definitions are given below. In either case, one needs to choose the three lattice vectors a_1 b_2, and a_3 as described above. Given these, the three primitive reciprocal lattice vectors are also determined (denoted b_1, b_2 and b_3).

Then, given the three Miller indices ℓ, m, n , (ℓmn) denotes planes orthogonal to:

$\ell b_1 + m b_2 + n b_3$.

That is, (ℓmn) simply indicates a normal to the planes in the basis of the primitive reciprocal lattice vectors. Because the coordinates are integers, this normal is itself always a reciprocal lattice vector. The requirement of lowest terms means that it is the *shortest* reciprocal lattice vector in the given direction.

Equivalently, (ℓmn) denotes a plane that intercepts the three points a_1/ℓ ,a_2/m and a_3/n or some multiple thereof. That is, the Miller indices are proportional to the *inverses* of the intercepts of the plane, in the basis of the lattice vectors. If one of the indices is zero, it means that the planes do not intersect that axis (the intercept is "at infinity").

The related notation $[\ell mn]$ denotes the *direction*:

$$\ell a_1 + m a_2 + n a_3$$

That is, it uses the direct lattice basis instead of the reciprocal lattice. Note that $[\ell mn]$ is *not* generally normal to the (ℓmn) planes, except in a cubic lattice.

CRYSTALLOGRAPHIC PLANES AND DIRECTIONS

The crystallographic directions are fictitious lines linking nodes (atoms, ions or molecules) of a crystal. The crystallographic planes are fictitious planes linking nodes. Some directions and planes have a higher density of nodes; these dense planes have an influence on the behaviour of the crystal:

- *Optical properties:* in condensed matter, the light "jumps" from one atom to the other with the Rayleigh scattering; the velocity of light thus varies according to the directions, whether the atoms are close or far; this gives the birefringence
- *Adsorption and reactivity:* the adsorption and the chemical reactions occur on atoms or molecules, these phenomena are thus sensitive to the density of nodes;
- *Surface tension:* the condensation of a material means that the atoms, ions or molecules are more stable if they are surrounded by other similar species; the surface tension of an interface thus varies according to the density on the surface
    - The pores and crystallites tend to have straight grain boundaries following dense planes
    - Cleavage
- Dislocations (plastic deformation)
    - The dislocation core tends to spread on dense planes (the elastic perturbation is "diluted"); this reduces the friction (Peierls-Nabarro force), the sliding occurs more frequently on dense planes;
    - The perturbation carried by the dislocation (Burgers vector) is along a dense direction: the shift of one node in a dense direction is a lesser distortion;
    - the dislocation line tends to follow a dense direction, the dislocation line is often a straight line, a dislocation loop is often a polygon.

For all these reasons, it is important to determine the planes and thus to have a notation system.

Case of the Cubic Structures

For the special case of cubic crystals, the lattice vectors are orthogonal and of equal length; similarly for the reciprocal lattice. So, in this common case, the Miller indices and both simply denote normals/directions in Cartesian coordinates.

Case of Hexagonal and rhombohedral Structures

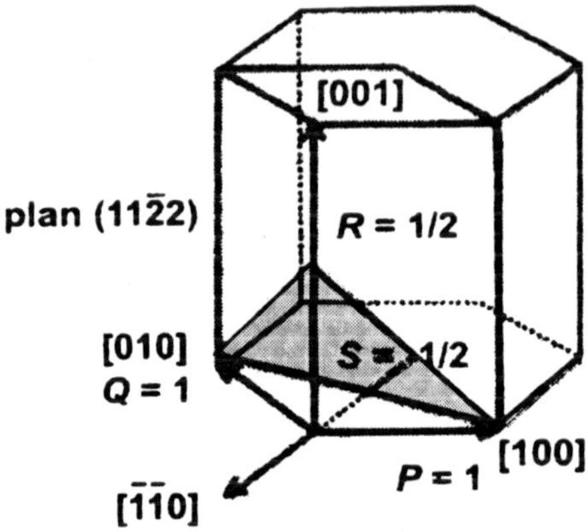

With hexagonal and rhombohedral crystal systems, it is possible to use the Bravais-Miller index which has 4 numbers $(h\ k\ i\ l)$

$i = -h-k$

where h, k and l are identical to the Miller index.

The (100) plane has a 3-fold symmetry, it remains unchanged by a rotation of 1/3 ($2\pi/3$ rad, 120°). The [100], [010] and the $(\bar{1}\bar{1}0)$ directions are really similar. If S is the intercept of the plane with the $(1\bar{1}0)$ axis, then

$i = 1/S$

i is redundant and not necessary.

- Coordinates in *square brackets* such as [100] denote a direction (in real space).
- Coordinates in *angle brackets* or *chevrons* such as <100> denote a *family* of directions which are equivalent due to symmetry operations. If it refers to a cubic system, this example could mean [100], [010], [001] or the negative of any of those directions.

- Miller indices in *parentheses* such as (100) denote a plane, in a cubic system the normal to the (hkl) planes the direction [hkl].
- Indices in *curly brackets* or *braces* such as {100} denote a family of plane normals which are equivalent due to symmetry operations, much the way angle brackets denote a family of directions.

TECHNIQUE

Some materials studied using crystallography, proteins for example, do not occur naturally as crystals. Typically, such molecules are placed in solution and allowed to crystallise over days, weeks, or months through vapour diffusion. A drop of solution containing the molecule, buffer, and precipitants is sealed in a container with a reservoir containing a hygroscopic solution. Water in the drop diffuses to the reservoir, slowly increasing the concentration and allowing a crystal to form. If the concentration were to rise more quickly, the molecule would simply precipitate out of solution, resulting in disorderly granules rather than an orderly and hence usable crystal.

Once a crystal is obtained, data can be collected using a beam of radiation. Although many universities that engage in crystallographic research have their own X-ray producing equipment, synchrotrons are often used as X-ray sources, because of the purer and more complete patterns such sources can generate. Synchrotron sources also have a much higher intensity of X-ray beams, so data collection takes a fraction of the time normally necessary at weaker sources.

Producing an image from a diffraction pattern requires sophisticated mathematics and often an iterative process of modelling and refinement. In this process, the mathematically predicted diffraction patterns of an hypothesised or "model" structure are compared to the actual pattern generated by the crystalline sample. Ideally, researchers make several initial guesses, which through refinement all converge on the same answer. Models are refined until their predicted patterns match to as great a degree as can be achieved without radical revision of the model. This is a painstaking process, made much easier today by computers.

The mathematical methods for the analysis of diffraction data only apply to *patterns,* which in turn result only when waves diffract from orderly arrays. Hence crystallography applies for the most part only to crystals, or to molecules which can be coaxed to crystalise for the sake of measurement. In spite of this, a certain amount of molecular information can be deduced from the patterns that are generated by fibres and powders, which while not as perfect as a solid crystal, may exhibit a degree of order. This level of order can be sufficient to deduce the structure of simple molecules, or to determine the coarse features of more complicated molecules (the double-helical structure of DNA, for example, was deduced from an X-ray diffraction pattern that had been generated by a fibrous sample).

Crystallography in Materials Engineering

Crystallography is a tool that is often employed by materials scientists. In single crystals, the effects of the crystalline arrangement of atoms is often easy to see macroscopically, because the natural shapes of crystals reflect the atomic structure. In addition, physical properties are often controlled by crystalline defects. The understanding of crystal structures is an important prerequisite for understanding crystallographic defects. Mostly, materials do not occur in a single crystalline, but poly-crystalline form, such that the powder diffraction method plays a most important role in structural determination.

A number of other physical properties are linked to crystallography. For example, the minerals in clay form small, flat, plate-like structures. Clay can be easily deformed because the plate-like particles can slip along each other in the plane of the plates, yet remain strongly connected in the direction perpendicular to the plates. Such mechanisms can be studied by crystallographic texture measurements.

In another example, iron transforms from a body-centred cubic (bcc) structure to a face-centred cubic (fcc) structure called austenite when it is heated. The fcc structure is a close-packed structure, and the bcc structure is not, which explains why the volume of the iron decreases when this transformation occurs.

Crystallography is useful in phase identification: That is, when performing some kind of processing on a material, it is often desired to find out what compounds and what phases are present in the material. Each phase has a characteristic arrangement of atoms. Techniques like X-ray diffraction can be used to identify which patterns are present in the material, and thus which compounds are present (note: the determination of the "phases" within a material should not be confused with the more general problem of "phase determination," which refers to the phase of waves as they diffract from planes within a crystal, and which is a necessary step in the interpretation of complicated diffraction patterns).

Crystallography covers the enumeration of the symmetry patterns which can be formed by atoms in a crystal and for this reason has a relation to group theory and geometry.

Crystallography is a fascinating division of the entire study of mineralogy. Even the non-collector may have an appreciation for large well-developed beautifully symmetrical individual crystals, like those of pyrite from Spain, and groups of crystals, such as quartz from Arkansas or tourmaline from California, because they are aesthetically pleasing. To think that such crystals come from the ground "as is" surprising to many. The lay person simply hasn't had the opportunity to learn about crystals and why they are the way they are; however, neither have many rockhounds and hobbyists.

We hope to bring you to a greater appreciation of natural mineral crystals and their forms by giving you some background and understanding into the world of crystallography.

Crystallography is simply a fancy word meaning "the study

of crystals". At one time the word crystal referred only to quartz crystal, but has taken on a broader definition which includes all minerals with well expressed crystal shapes.

Crystallography may be studied on many levels, but no matter how elementary or in-depth a discussion of the topic we have, we confront some geometry.

Crystallography is easily divided into 3 sections — geometrical, physical, and chemical. The latter two involve the relationships of the crystal form (geometrical) upon the physical and chemical properties of any given mineral. We will cover the most significant geometric aspects of crystallography.

In any type of study, there exists special words used to summarise entire concepts. This is the special language of the "expert", whether you speak of electrical engineering, computer science, accounting, or crystallography. There's no real way to get around learning some of these basic definitions and "laws" so we might as well jump right into them.

First, let's define what we're dealing with. A Crystal is a regular polyhedral form, bounded by smooth faces, which is assumed by a chemical compound, due to the action of its interatomic forces, when passing, under suitable conditions, from the state of a liquid or gas to that of a solid. Let's dissect that statement. A polyhedral form simply means a solid bounded by flat planes (we call these flat planes Crystal Faces). "A chemical compound" tells us that all minerals are chemicals, just formed by and found in nature. The last half of the definition tells us that a crystal normally forms during the change of matter from liquid or gas to the solid state. In the liquid and gaseous state of any compound, the atomic forces that bind the mass together in the solid state are not present. Therefore, we must first crystallise the compound before we can study it's geometry. Liquids and gases take on the shape of their container, solids take on one of several regular geometric forms. These forms may be subdivided, using geometry, into six systems.

But before we can begin to discuss the individual systems and their variations, let's address several other topics which we will use to describe the crystal systems. There's also some laws and rules we must learn.

Way back in 1669, Nicholas Steno, a Danish physician and natural scientist, discovered one of these laws. By examination of numerous specimens of the same mineral, he found that, when measured at the same temperature, the angles between similar crystal faces remain constant regardless of the size or the shape of the crystal. So whether the crystal grew under ideal conditions or not, if you compare the angles between corresponding faces on various crystals of the same mineral, the angle remains the same.

Although he did not know why this was true (x-rays had not yet been discovered, much less x-ray diffraction invented), we now know that this is so because studies of the atomic structure of any mineral proves that the structure remains within a close set of given limits or geometric relationships. If it doesn't, then by the modern definition of a mineral, we are not comparing the same two minerals. We might be comparing polymorphs, but certainly not the same mineral! (Polymorphs being minerals with the same chemistry, like diamond and graphite or sphalerite and wurtzite, but having differing atomic structure and, therefore, crystallising in different crystal systems). Steno's law is called the (Constancy Of Interfacial Angles) and, like other laws of physics and chemistry, we just can't get away from it.

Now, some of you may be thinking: I have a mineral crystal that does not match the pictures in the mineral books. What you may have is a distorted crystal form where some faces may be extremely subordinate or even missing. Distorted crystals are common and result from less-than-ideal growth conditions or even breakage and recrystallisation of the mineral. However, remember that we must also be comparing the angles between similar faces. If the faces are not present, then you cannot compare them. With many crystals we are dealing with a final shape determined by forces other than those of the interatomic bonding.

During the process of crystallisation in the proper environment, crystals assume various geometric shapes dependent on the ordering of their atomic structure and the physical and chemical conditions under which they grow. If there is a predominant direction or plane in which the mineral forms, different habits prevail. Thus, galena often forms equate shapes (cubes or octahedrons), quartz typically is prismatic, and barite tabular.

To discuss the six crystal systems, we have to establish some understanding of solid geometry. To do this, we will define and describe what are called 'Crystallographic Axes'. Since we are dealing with 3 dimensions, we must have 3 axes and, for the initial discussion, let's make them all equal and at right angles to each other. This is the simplest case to consider. The axes pass through the centre of the crystal and, by using them, we can describe the intersection of any given face with these 3 axes.

Mineralogists had to decide what to call each of these axes and what their orientation in each crystal was so that everyone was talking the same language. Many different systems arose in the early literature. Then, as certain systems were found to have problems, some were abandoned until we arrived at the notational system used today. There exists two of these presently in use, complimentary to each other. One uses number notation to indicate forms or individual faces and the other uses letters to indicate forms. But let's get back to our 3 axes and we'll discuss these two systems later.

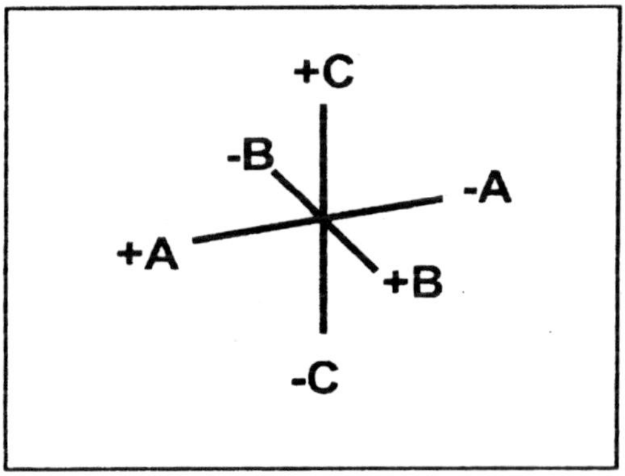

We are going to draw each axis on a sheet of paper and describe its orientation. All you need for this exercise is a pencil and paper. Make the first axis vertical, and we'll call it the c axis. The top is the + end and the bottom is the - end. The second axis, the b axis, is horizontal and passes through the centre of the c axis. It is the same length as the c axis. The right end is the +, and the left is the -. The third axis is the a axis and passes at a right angle through the join of the b and c axes.

It is somewhat tricky to draw because, even though the a axis is the same length as the c and b axes, because it goes from front to back it appears shorter. It is hard to represent a 3-dimensional figure on the 2-dimensional surface of paper, but you can do it.

You have to use a sense of perspective, as an artist would say. The front end of a, which appears to come out of the paper, is the + and the back end the - (appears to be in the background or behind the paper).

Perspective is a key to drawing 3 dimensional objects on a flat 2D piece of paper. Perspective is what makes railroad tracks look like they come together in the distance. It is also what causes optical illusions when trying to draw axial crosses, or line drawings of crystal models.

Perhaps you've looked at these lines and tried to decide which one comes forward, and which one recedes, and then have the illusion flip-flop so that it looks the other way. That's why they are labelled + and -. It can be confusing, so don't feel like the lone ranger.

We have now reached the point in our discussion that we can actually mention the six large groups or crystal systems that all crystal forms may be placed in. We will use our crystallographic axes which we just discussed to subdivide all known minerals into these systems. The systems are:

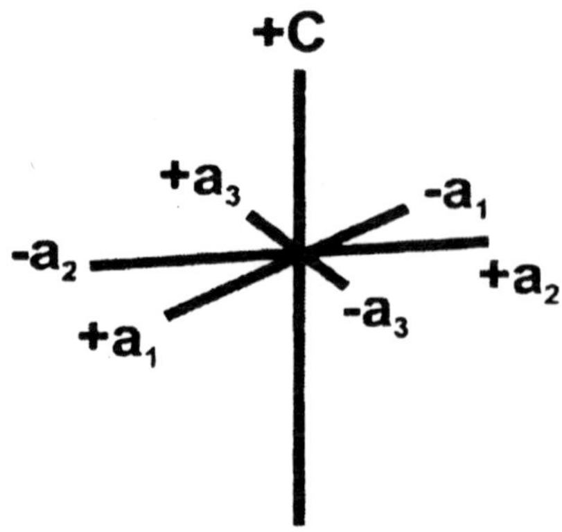

Figure

(1) *Cubic (Aka Isometric):* The three crystallographic axes are all equal in length and intersect at right angles (90 degrees) to each other. However, we now will rename the axes a_1, a_2, and a_3 because they are the same length (a becomes a1, b becomes a_2, and c becomes a_3).

(2) *Tetragonal:* Three axes, all at right angles, two of which are equal in length (a and b) and one (c) which is different in length (shorter or longer). Note: If c was equal in length to a or b, then we would be in the cubic system.

(3) *Orthorhombic:* Three axes, all at right angles, and all three of different lengths. Note: If any axis was of equal length to any other, then we would be in the tetragonal system.

(4) *Hexagonal:* Four axes! We must define this situation since it can not be derived. Three of the axes fall in the same plane and intersect at the axial cross at 120 degrees between the positive ends. These 3 axes, labelled a_1, a_2, and a_3, are the same length. The fourth axis, termed c, may be longer or shorter than the a axes set. The c axis also passes through the intersection of the a axes set at right angle to the plane formed by the a set.

(5) *Monoclinic:* Three axes, all unequal in length, two of which (a and c) intersect at an oblique angle (not

90 degrees), the third axis (b) is perpendicular to the other two axes. Note: If a and c crossed at 90 degrees, then we would be in the orthorhombic system.

(6) *Triclinic:* The three axes are all unequal in length and intersect at three different angles (any angle but 90 degrees). Note: If any two axes crossed at 90 degrees, then we would be describing a monoclinic crystal.

As was stated earlier, all known crystal forms fit into the above six crystal systems. But why don't all crystals in a given set look the same? Or, stated differently, why can't I learn six crystal shapes and know all I need to know? Well, crystals, even of the same mineral, have differing Crystal Forms, depending upon their conditions of growth. Whether they grew rapidly or slowly, under constant or fluctuating conditions of temperature and pressure, or from highly variable or remarkably uniform fluids or melts, all these factors have their influence on the resultant crystal shapes, even when not considering other controls.

We must touch on a type of notation often seen in mineral literature known as Miller Indices. Before William H. Miller (1801-1880) devised this mathematical system for describing any crystal face or group of similar faces (forms), there existed a considerable amount of confusion due to the many different descriptive systems. Some of these systems used letter symbols to denote crystal faces and forms. Also different mineralogical "schools" existed as to how a given crystal should be viewed or oriented before assigning the crystallographic axes and then describing the various faces and forms present. If you were of the German school, you had one view; from the English school another thought; from the French school, still another opinion.

So the problem was really one of how to bring order to the literature's chaos. To the problem, Miller (University of Cambridge) applied relatively simple mathematics - the Universal Language. To the lettering systems, Miller described the a,b,c (for hexagonal crystals his notation is four numbers long) intercepts of each planar crystal form as numbers and also made note of the form letter. His numbering system became widely accepted and is known as Miller indices. The numbers are presented as whole numbers (fractions are not allowed) and are the reciprocal of the actual intercept number, all whole numbers being reduced by their lowest common denominator. Here's a couple of simple examples from the cubic system.

Let us first describe a face of an octahedron and later a cube using Miller's indices. First, we should realise that an octahedron is an eight-sided crystal form that is the simple repetition of an equilateral triangle about our 3 crystallographic axes. The triangle is oriented so that it crosses the a_1, a_2, and a_3 axes all at the same distance from the axial cross. This unit distance is given as 1. Dividing 1 into the whole number 1 (it's a reciprocal, remember?) yields a value of 1 for each Miller number. So the Miller indices is (111) for the face that intercepts the positive end of each of the 3 axes. Note: A bar over the number tells us that the intercept was across the negative end of the particular crystallographic axis. The octahedral form is given the letter designation of "o".

Now to the cube face. A cube face that intercepts the a_3 (vertical) axis on the + end will not intercept the a_1 and a_2 axes. If the face does not intercept an axis, then we assign a mathematical value of infinity to it. So we start with Infinity, Infinity, 1 (a_1, a_2, a_3). Infinity divided into 0 = 0 (any number divided into zero equals zero). So the Miller indices of the $+a_3$ intercept face equals (001).

We should briefly mention cleavage at this point. Cleavage is the preferred planar direction of breakage that many minerals possess. It is due to planes of weakness that exist in some minerals because the bonding strength between the atoms or molecules is not the same in every direction. Because crystals are composed of orderly arrangements of atoms or molecules, we really should expect cleavage to be present in many crystals. The notation that denotes cleavage is derived in much the same manner as Miller indices, but is expressed in braces. So a cubic crystal, say diamond, no matter whether it exhibits cubic {001} or octahedral {111} crystal form, has an octahedral cleavage form that is given as {111}.

Note: The Miller indices when used as face symbols are enclosed in parentheses, as the (111) face for example. Form symbols are enclosed in braces, as the {111} form for example. Zone symbols are enclosed in brackets, [111] for example and denote a zone axis in the crystal. So in the discussion of cleavage you must use braces to denote cleavage. Cleavage is analogous to form as cubic, octahedral, or pinacoidal cleavage and does not refer to just one face of a form.

Now we are ready to discuss elements of symmetry. These include planes of symmetry, axes of symmetry, and centre of symmetry. These symmetry elements may be or may not be combined in the same crystal. Indeed, we will find that one crystal class or system has only one of these elements!

Any two dimensional surface (we can call it flat) that, when passed through the centre of the crystal, divides it into two symmetrical parts that are mirror images is a plane of symmetry. Any plane of symmetry divides the crystal form into two mirror images. Planes of symmetry are often referred to as mirror image planes. Let's discuss a cube again. A cube has 9 planes of symmetry, 3 of one set and 6 of another. We must use two figures to easily recognise all of them.

Planes of symmetry are always possible crystal forms. This means that, although not always present on many natural crystals, there exists the possibility that other crystal faces may be expressed. So even though a cube form does not present an octahedral face, it is always possible that it could have formed under the right conditions.

The typical human has two hands, right and left. Place them together palms facing away from you and the tips of the thumbs touching. Assuming that you have the same number of fingers on each hand, you will note that your right hand is the mirror image of your left and vise versa. The average person is symmetrical, having binary symmetry vertically from the head to the feet when viewed from the front or back (bilateral symmetry).

You can have a lot of laughs with friends and a long mirror using this symmetry element. You need at least one other person to do this so you both can view the results!.

It is sometimes convenient to designate planes of symmetry as axial, diagonal, principle, or intermediate.

Axes of symmetry can be rather confusing at first, but let's have a go at them anyway. Any line through the centre of the crystal around which the crystal may be rotated so that after a definite angular revolution the crystal form appears the same as before is termed an axis of symmetry. Depending on the amount or degrees of rotation necessary, four types of axes of symmetry are possible when you are considering crystallography (some textbooks list five). Given below are all possible rotational axes:

- When rotation repeats form every 60 degrees, then we have sixfold or hexagonal symmetry. A filled hexagon symbol is noted on the rotational axis.
- When rotation repeats form every 90 degrees, then we have fourfold or tetragonal symmetry. A filled square is noted on the rotational axis.
- When rotation repeats form every 120 degrees, then we have threefold or trigonal symmetry. A filled equilateral triangle is noted on the rotational axis.
- When rotation repeats form every 180 degrees, then we have twofold or binary symmetry. A filled oval is noted on the rotational axis.
- When rotation repeats form every 360 degrees, then we use a filled circle as notation. This one we consider optional to list as almost any object has this symmetry. If you really want to know the truth, this means no symmetry.

Note that rotational axes may be on the plane of the face,

on the edge of where two faces meet, or on the point of conjunction of three or more faces. On a complete crystal form, the axis must pass through the centre of the crystal and exist at the equivalent site on the opposite side of the crystal as it entered.

Take a solid cube, made of wood or plastic (a clear plastic cube box works well for this exercise). Mark, using the rotational notation, every four-, three-, and two-fold axis of rotation that you can find.

First, draw a crystal. Use a piece of "2 by 2" board and make this crystal form by cutting off the ends so the wood block looks like the drawing. Hold the block in your left hand with your thumb on the top and in the centre of the 2-face edge join (long axis) and your index finger on the same join on the underside. Your palm will be towards your body. Align your two fingers so that you are looking straight down on your thumb and can not see end of your index finger. The top of the block will appear as 2 equal-sized faces, sloping away from you. If you rotate the block 180 degrees, the faces will be appear back in the same position (2-fold axis of rotation), but here's the tricky part — rotate the specimen 90 degrees and then turn your wrist where your index finger is on top (easiest done by turning your wrist counterclockwise). You will see that the block's faces appear in the original position in the original position. You have discovered an axis of rotary inversion!

Depending upon what elements of symmetry are present, all crystals may be divided into 32 distinct groups called classes of symmetry. Remember, we concern ourselves with the symmetry elements we learned above, not the malformed crystal shapes of most minerals. Only forms which belong to the same class can occur in combination together in nature. We can not find a cube face on a hexagonal crystal. Likewise, we will never discover the rhombic dipyramid termination of a hexagonal crystal on a tetragonal crystal. So our laws, rules, and symmetry elements prevent chaos in our beautifully symmetrical world of crystallography. Certainly, when dealing with real crystals, distortion problems can arise! Think of capillary pyrite. Here you have a cubic crystal which, due to a growth phenomenon, has one axis nearing infinity in length in relation to the other two. But this is caused by special conditions during growth, not the crystallography.

There are graphical methods of plotting all possible crystal faces and symmetry elements on a type of diagram called a stereo net. Stereo nets give a way to represent three-dimensional data on a two-dimensional surface (a flat sheet of paper).

Almost all minerals are crystalline (part of the definition of a mineral), but only under special conditions do you form minerals with well expressed crystal form(s). The mineral may grow from a melt as in igneous rocks, from solids under elevated temperature and pressure, or in open crevices or pockets from hot, warm, or cold mineral-laden fluids.

So, when you examine euhedral crystals of any mineral, please realise that they are rather remarkable objects. Probably 99 per cent of all minerals present in the earth's crust display no external crystalline form. A euhedral crystal is a flower of the mineral world! Treat crystals with care and respect because they probably survived millions of years before someone discovered them. One careless moment and they may be destroyed forever.

Electron Crystallography

Electron crystallography is a method to determine the arrangement of atoms in solids using an electron microscope. It can complement X-ray crystallography on proteins (such as membrane proteins), that cannot easily form the large 3-dimensional crystals required for that process. Structures are usually determined from either 2-dimensional crystals (sheets or helices), polyhedrons such as viral capsids, or dispersed individual proteins. Electrons can be used in these situations, whereas X-rays cannot, because electrons interact more strongly with atoms than X-rays do. Thus, X-rays will travel through a thin 2-dimensional crystal without diffracting significantly, whereas electrons can be used to form an image. Conversely, the strong interaction between electrons and proteins makes thick (e.g. 3-dimensional) crystals impervious to electrons, which only penetrate short distances.

One of the main difficulties in X-ray crystallography is determining phases in the diffraction pattern. Because no X-ray lens exists, X-rays cannot be used to form an image of the crystal being diffracted, and hence phase information is lost. Fortunately, electron microscopes contain electron lenses, and phase information tends to be much more reliable in electron crystallography.

A common problem to X-ray crystallography and electron crystallography is radiation damage, by which proteins are damaged as they are being imaged, limiting the resolution that can be obtained. This is especially troublesome in the setting of electron crystallography, where that radiation damage is focused on far fewer atoms. One technique used to limit radiation damage is electron cryomicroscopy, in which the samples undergo cryofixation and imaging takes place at liquid nitrogen or even liquid helium temperatures. Because of this problem, X-ray crystallography has been much more successful in determining the structure of proteins that are especially vulnerable to radiation damage.

The first electron crystallographic protein structure to achieve atomic resolution was bacteriorhodopsin, determined by Richard Henderson and co-workers at the Medical Research Council Laboratory of Molecular Biology in 1990. Since then, several other high-resolution structures have been determined by electron crystallography, including the light-harvesting complex, the nicotinic acetylcholine receptor, and the bacterial flagellum.

Powder Diffraction

Powder diffraction is a scientific technique using X-Ray or neutron diffraction on powder or microcrystalline samples for structural characterisation of materials. Ideally, every possible crystalline orientation is represented equally in the sample, leading to smooth diffraction rings around the beam axis rather than the discrete Laue spots observed for single crystal diffraction. In accordance with Bragg's law, each ring corresponds to a particular reciprocal lattice vector in the sample crystal, with intensity proportional to the number of such planes. In practice, it is sometimes necessary to rotate the sample orientation to eliminate the effects of texturing and achieve true randomness. The dedicated machine to perform such measurements is called a powder diffractometer.

Advantages and Disadvantages

The great advantages of the technique are simplicity of sample preparation, rapidity of measurement, a better representation of some real materials, and the ability to analyse mixed phases. Growth and mounting of large single crystals is notoriously difficult, whereas many materials are readily available with sufficient microcrystallinity for powder diffraction, or samples may be easily ground from larger crystals. Particularly for neutron diffraction, which requires larger samples than X-Ray Diffraction due to a relatively weak scattering cross-section, the ability to prepare large samples can be critical.

Since all possible crystal orientations are measured simultaneously, collection times can be quite short even for small and weakly scattering samples. This is not merely convenient, but can be essential for samples which are unstable either inherently or under x-ray or neutron bombardment, or for time-resolved studies. As with any technique, sample preparation must be taken into account before analysis. For instance, the crystalline grains in industrial rolled steel strongly orient with the more compressible hexagonal c-axis perpendicular to the plane of rolling. This texturing distorts the observed peak intensity ratios despite a grain size which would otherwise be sufficiently small for the powder condition to obtain. In this case, sample rotation is required for accurate crystal phase determination. Once the phase is known, the degree of preferential orientation may be determined for various preparations by directly measuring the degree of texturing, which information is never available through single crystal diffraction. Mixed phases occur when multiple crystalline arrangements of scatterers are thermodynamically stable or metastable. For instance, tin exhibits two allotropes at ambient temperature and pressure. At low temperatures, the a-Sn, or grey tin, phase is favoured, and the degree to which a sample has converted may be measured with powder diffraction.

Powder diffraction data is measured in terms of reciprocal space, analysis of which is less straightforward than single crystal data. Further, the three dimensional information of the reciprocal lattice of a crystal is collapsed into a one dimensional diffractogram, yielding somewhat less information than a corresponding single crystal diffraction experiment.

Uses

Relative to other methods of analysis, powder diffraction allows for rapid, non-destructive analysis of multi-component mixtures without the need for extensive sample preparation. This gives laboratories around the world the ability to quickly analyse unknown materials and perform materials characterisation in such fields as metallurgy, mineralogy, forensic science, archeology and the biological and pharmaceutical sciences. Advances in hardware and software, particularly improved optics and fast detectors, have dramatically improved the analytical capability of the technique, especially relative to the speed of the analysis. The fundamental physics upon which the technique is based provides high precision and accuracy in the measurement of interplanar spacings, sometimes to fractions of an Angstrom, resulting in authoritative identification frequently used in patents, criminal cases and other areas of law enforcement. The ability to analyse multiphase materials also allows analysis of how materials interact in a particular matrix such as a pharmaceutical tablet, a circuit board, a mechanical weld, a geologic core sampling, cement and concrete, or a pigment found in an historic painting. The method has been historically used for the identification and classification of minerals, but it can be used for any materials, even amorphous ones, so long as a suitable reference pattern is known or can be constructed.

Crystal Structure Determination

The most widespread use of powder diffraction is in the identification and characterisation of crystalline solids, each of which produces a distinctive diffraction pattern. Both the positions (corresponding to lattice spacings) and the relative intensity of the lines are indicative of a particular phase and material, providing a "fingerprint" for comparison. A multi-phase mixture, e.g. a soil sample, will show more than one pattern superposed, allowing for determination of relative concentration.

J.D. Hanawalt, an Analytical Chemist who worked for Dow Chemical in the 1930s, was the first to realise the analytical potential of creating a database. Today it is represented by the Powder Diffraction File (PDF) of the International Centre for Diffraction Data (formerly Joint Committee for Powder Diffraction Studies). This has been made searchable by computer through the work of global software developers and equipment manufacturers. There are now over 550,000 reference materials in the 2006 Powder Diffraction File Databases, and these databases are interfaced to a wide variety of diffraction analysis software and distributed globally. The Powder Diffraction File contains many subfiles, such as minerals, metals and alloys, pharmaceuticals, forensics, excipients, superconductors, semi-conductors, etc., with large collections of organic, organometallic and inorganic reference materials.

Crystallinity

In contrast to a crystalline pattern consisting of a series of sharp peaks, amorphous materials (liquids, glasses, etc.) produce a broad background signal. Many polymers show semi-crystalline behaviour, *i.e.* part of the material forms an ordered crystallite by folding of the molecule. One and the same molecule may well be folded into two different crystallites and thus form a tie between the two. The tie part is prevented from crystallising. The result is that the crystallinity will never reach 100 per cent. Powder XRD can be used to determine the crystallinity by comparing the integrated intensity of the background pattern to that of the sharp peaks.

Crystallite Size

If the crystallites of the powder are very small the peaks of the pattern will broaden. From the broadening it is possible to determine an average crystallite size, in A, by Debye-Scherrer formula: $D_{hkl} = \frac{\kappa\tau}{\beta \cos\theta}$

here κ = 0.8 — 1.39 (usually close to unity, e.g. 0.9), λ wavelength of the radiation λ_{Cu} = 1.54056 A, β - FWHM (full width at half maximum, or half-width) in radians, β = half-width (degree) Pi/180, θ - the position of the maximum of diffraction.

Note: on the diffractogram you usually plot 2θ on the x axis, not θ.

An error for the crystallite size by this formula can be up to 50 per cent, so caution is needed when making assertions based solely on this technique.

Phase Transitions

Powder diffraction can be combined with *in situ* temperature and pressure control. As these thermodynamic variables are changed, the observed diffraction peaks will migrate continuously to indicate higher or lower lattice spacings as the unit cell distorts. This allows for measurement of such quantities as the thermal expansion tensor and the isothermal bulk modulus, as well determination of the full equation of state of the material. At some critical set of conditions, for instance 0 °C for water at 1 atm, a new arrangement of atoms or molecules may become stable, leading to a phase transition. At this point new diffraction peaks will appear or old ones disappear according to the symmetry of the new phase.

Magnetic Structures and Detection of Hydrogen

X-ray photons scatter by interaction with the electron cloud of the material, neutrons are scattered by the nuclei. This means that the scattering power for XRD is roughly linear with atomic number, but fluctuates from isotope to isotope for neutrons. Both protons and deuterons (the stable nuclei of the element hydrogen) are strong scatterers for neutrons. The one electron of the atom makes it difficult to detect by X-rays. This makes neutron powder diffraction an attractive way to gain better insight of the precise location of hydrogen in a structure.

Neutrons can also undergo scattering if the material has an ordered magnetic structure. Neutrons have a spin and thus a small magnetic moment which interacts with the magnetic moments of electrons in an incomplete shell. The powder pattern can be used to determine the magnetic structure.

Two-dimensional powder diffraction setup for high energy X-rays. HEX-Rays entering from the left are diffracted in forward direction at the sample and registered by a 2D detector such as an image plate.

Cameras

The simplest cameras for X-ray powder diffraction consist of a small capillary and either a flat plate detector (originally a piece of X-ray film, now more and more a flat-plate detector or a CCD-camera) or a cylindrical one (originally a piece of film in a cookie-jar, now more and more a bent position sensitive detector). The two types of cameras are known as the Laue and the Debye-Scherrer camera.

To promote randomisation the capillary is usually spinning around its axis.

For neutron diffraction vanadium cylinders are used as sample holders. Vanadium is all but transparent for neutrons. The element hardly scatters at all.

A later development in X-ray cameras is the Guinier camera. It is built around a *focusing* bent crystal monochromator. The sample is usually placed in the focusing beam, e.g. as a dusting on a piece of sticky tape. A cylindrical piece of film (or electronic multichannel detector) is put on the focusing circle, but the incident beam prevented from reaching the detector to prevent damage from its high intensity.

Diffractometers

Diffractometers can be operated both in transmission and in reflection configurations. The reflection is more common. The powder sample is filled in a small disc like container and its surface carefully flattened. The disc is put on one axis of the diffractometer and tilted by an angle θ while a detector (scintillation counter) rotates around it on an arm at twice this angle. This configuration is known under the name Bragg-Brentano. The availability of position sensitive detectors and CCD-cameras is making this type of equipment more and more obsolete.

Neutron Diffraction

Neutron sources suitable for diffraction are only available at a small number of research reactors and spallation sources in the world. To compensate for the low flux of neutrons, they usually have a battery of individual detectors arranged in a cylindrical fashion around the sample holder, and can therefore collect scattered intensity simultaneously on a large 2θ range.

5

Crystal Structure and Crystal System

The discipline of crystallography has developed a descriptive terminology which is applied to crystals and crystal features in order to describe their structure, symmetry, and shape. This terminology defines the crystal lattice which provides a mineral with its ordered internal structure. It also describes various types of symmetry. By considering what type of symmetry a mineral species possesses, the species may be categorised as a member of one of six crystal systems and one of thirty-two crystal classes.

The concept of symmetry describes the periodic repetition of structural features. Two general types of symmetry exist. These include translational symmetry and point symmetry. Translational symmetry describes the periodic repetition of a motif across a length or through an area or volume. Point symmetry, on the other hand, describes the periodic repetition of a motif around a point.

Reflection, rotation, inversion, and rotoinversion are all

point symmetry operations. A reflection occurs when a motif on one side of a plane passing through the centre of a crystal is the mirror image of a motif which appears on the other side of the plane. The motif is said to be reflected across the mirror plane which divides the crystal. Rotational symmetry arises when a structural element is rotated a fixed number of degrees about a central point before it is repeated. If a crystal possesses inversion symmetry, then every line drawn through the centre of the crystal will connect two identical features on opposite sides of the crystal. Rotoinversion is a compound symmetry operation which is produced by performing a rotation followed by an inversion.

A specified motif which is translated linearly and repeated many times will produce a lattice. A lattice is an array of points which define a repeated spatial entity called a unit cell. The unit cell of a lattice is the smallest unit which can be repeated in three dimensions in order to construct the lattice. The corners of the unit cell serve as points which are repeated to form the lattice array; these points are termed lattice points.

The number of possible lattices is limited. In the plane only five different lattices may be produced by translation. The French crystallographer Auguste Bravais (1811-1863) established that in three-dimensional space only fourteen different lattices may be constructed. These fourteen different lattice structures are thus termed the Bravais lattices.

The reflection, rotation, inversion, and rotoinversion symmetry operations may be combined in a variety of different ways. There are thirty-two possible unique combinations of symmetry operations. Minerals possessing the different combinations are therefore categorised as members of thirty-two crystal classes; each crystal class corresponds to a unique set of symmetry operations. Each of the crystal classes is named according to the variant of a crystal form which it displays. Each crystal class is grouped as one of the six different crystal systems according to which characteristic symmetry operation it possesses.

A crystal form is a set of planar faces which are geometrically equivalent and whose spatial positions are related to one another by a specified set of symmetry operations. If one face of a crystal form is defined, the specified set of point symmetry operations will determine all of the other faces of the crystal form.

A simple crystal may consist of only a single crystal form. A more complicated crystal may be a combination of several different forms. The crystal forms of the five non-isometric crystal systems are the monohedron or pedion, parallelohedron or pinacoid, dihedron, or dome and sphenoid, disphenoid, prism, pyramid, dipyramid, trapezohedron, scalenohedron, rhombohedron and tetrahedron. Fifteen different forms are possible within the isometric system.

Each crystal class is a member of one of six crystal systems. These systems include the isometric, hexagonal, tetragonal, orthorhombic, monoclinic, and triclinic systems. The hexagonal crystal system is further broken down into the hexagonal and rhombohedral divisions. Every crystal of a certain crystal system will share a characteristic symmetry element with the other members of its system. The crystal system of a mineral species may sometimes be determined visually by examining a particularly well-formed crystal of the species.

SYMMETRY

Crystals possess a regular, repetitive internal structure. The concept of symmetry describes the repetition of structural features. Crystals therefore possess symmetry, and much of the discipline of crystallography is concerned with describing and cataloguing different types of symmetry.

Two general types of symmetry exist. These consist of translational symmetry and point symmetry. Translational symmetry describes the periodic repetition of a structural feature across a length or through an area or volume. Point symmetry, on the other hand, describes the periodic repetition of a structural feature around a point. Reflection, rotation, and inversion are all point symmetries.

Lattices

The concept of a lattice is directly related to the idea of translational symmetry. A lattice is a network or array composed of single motif which has been translated and repeated at fixed intervals throughout space. For example, a square which is translated and repeated many times across the plane will produce a planar square lattice.

The unit cell of a lattice is the smallest unit which can be repeated in three dimensions in order to construct the lattice. In a crystal, the unit cell consists of a specific group of atoms which are bonded to one another in a set geometrical arrangement. This unit and its constituent atoms are then repeated over and over in order to construct the crystal lattice. The surroundings in any given direction of one corner of a unit cell must be identical to the surroundings in the same direction of all the other corners. The corners of the unit cell therefore serve as points which are repeated to form a lattice array; these points are termed lattice points. The vectors which connect a straight line of equivalent lattice points and delineate the edges of the unit cell are known as the crystallographic axes.

The number of possible lattices is limited. In the plane only five different lattices may be produced by translation. One of these lattices possesses a square unit cell while another possesses a rectangular unit cell. The third possible planar lattice possesses a centred rectangular unit cell, which contains a lattice point in the centre as well as lattice points on the corners. The unit cell of the fourth possible planar lattice is a parallelogram, and that of the final planar lattice is a hexagonal unit cell which may alternately be considered a rhombus.

Bravais Lattices

The French crystallographer Auguste Bravais (1811-1863) established that in three-dimensional space only fourteen different lattices may be constructed. The fourteen Bravais lattices may be divided among six crystal systems. These are the isometric or cubic, tetragonal, orthorhombic, monoclinic, triclinic, and hexagonal systems. The Bravais lattices are furthermore of three different types. A primitive lattice has only a lattice point at each corner of the three-dimensional unit cell. A body-centred lattice contains not only lattice points at each corner of the unit cell but also contains a lattice point at the centre of the three-dimensional unit cell. A face-centred lattice possesses not only lattice points at the corners of the unit cell but also at either the centres of just one pair of faces or else at the centres of all three pairs of faces. The fourteen Bravais lattices are therefore the primitive cubic, body-centred cubic, face-centred cubic, primitive tetragonal, body-centred tetragonal, primitive orthorhombic, body-centred orthorhombic, single face-centred orthorhombic, multiple face-centred orthorhombic, primitive monoclinic, single face-centred monoclinic, primitive triclinic, single face-centred hexagonal, and rhombohedral lattices. (The rhombohedral lattice is a subset of the hexagonal crystal system.)

Point Symmetry Operations

Point symmetry describes the repetition of a motif or structural feature around a single reference point, commonly the centre of a unit cell or a crystal. The different point-symmetry operations are reflection, rotation, inversion, and the combined operation rotoinversion.

A reflection occurs when the structure features on one side of a plane passing through the centre of a crystal are the mirror image of the structural features on the other side. The plane across which the reflection occurs is then termed a mirror plane.

Rotational symmetry arises when a structural element is rotated a fixed number of degrees about a central point and then repeated. A square, for example, possesses 4-fold rotational symmetry because it may be rotated four times by 90° about its central point before it is returned to its original position. Each time it is rotated by 90° the resultant square will be identical in appearance to the original square.

If a crystal possesses inversion symmetry, then any line which is drawn through the origin at the centre of the crystal will connect two identical features on opposite sides of the crystal.

Rotoinversion is a compound symmetry operation which is produced by performing a rotation followed by an inversion. 1-fold, 2-fold, 3-fold, 4-fold, and 6-fold rotoinversion operations exist. Most of these rotoinversions may alternately be described by a specified set of rotation, reflection and inversion operations. A 1-fold rotoinversion is equivalent to rotation by 360° followed by inversion. This procedure is ultimately equivalent to a single inversion. A 2-fold rotoinversion axis is equivalent to reflection through a mirror plane perpendicular to the rotoinversion axis.

A crystal which possesses a 3-fold rotoinversion axis is equivalent to one which possesses both 3-fold rotational symmetry and inversion symmetry. A 6-fold rotoinversion is equivalent to 3-fold rotation and reflection across a mirror plane which lies at right angles to the rotation axis. The only rotoinversion operation which cannot be replaced by a combination of rotations, reflections and inversions is 4-fold rotoinversion.

The reflection, rotation, inversion, and rotoinversion symmetry operations may be combined in a variety of different ways. There are thirty-two different possible combinations of these symmetry elements. Minerals possessing the different combinations are therefore categorised as members of 32 possible crystal classes. According to this schema, each crystal class corresponds to a unique set of symmetry operations. Each crystal class is then placed into one of the six different crystal systems so that several different classes are members of each system.

Crystal Systems

Every crystal class is a member of one of the six crystal systems. These systems include the isometric, hexagonal, tetragonal, orthorhombic, monoclinic, and triclinic crystal systems. The hexagonal crystal system is further broken down into hexagonal and rhombohedral divisions.

Every crystal class which belongs to a certain crystal system will share a characteristic symmetry element with the other members of its system. For example, all crystals of the isometric system possess four 3-fold axes of symmetry which proceed diagonally from corner to corner through the centre of the cubic unit cell. In contrast, all crystals of the hexagonal division of the hexagonal system possess a single six-fold axis of rotation.

In addition to the characteristic symmetry element, a crystal class may possess other symmetry elements which are not necessarily present in all members of the same system. The crystal class which possesses the highest possible symmetry or the highest number of symmetry elements within each system is termed the holomorphic class of the system. For example, crystals of the holomorphic class of the isometric system possess inversion symmetry, three 4-fold axes of rotational symmetry, the characteristic set of four 3-fold axes of rotational symmetry which is indicative of the isometric crystal system, six 2-fold axes of rotational symmetry, and nine different mirror planes. In contrast, a crystal which is not a member of the holomorphic class yet still belongs to the isometric system may possess only three 2-fold axes of rotational symmetry and the characteristic four 3-fold axes of rotational symmetry.

The crystal system of a mineral species may sometimes be determined in the field by visually examining a particularly well-formed crystal of the species.

Isometric

The isometric crystal system is also known as the cubic system. The crystallographic axes used in this system are of equal length and are mutually perpendicular, occurring at right angles to one another.

All crystals of the isometric system possess four 3-fold axes of symmetry, each of which proceeds diagonally from corner to corner through the centre of the cubic unit cell. Crystals of the isometric system may also demonstrate up to three separate 4-fold axes of rotational symmetry. These axes, if present, proceed from the centre of each face through the origin to the centre of the opposite face and correspond to the crystallographic axes. Furthermore crystals of the isometric system may possess six 2-fold axes of symmetry which extend from the centre of each edge of the crystal through the origin to the centre of the opposite edge. Minerals of this system may demonstrate up to nine different mirror planes.

Examples of minerals which crystallise in the isometric system are halite, magnetite, and garnet. Minerals of this system tend to produce crystals of equidimensional or equant habit.

Hexagonal

Minerals of the hexagonal crystal system are referred to three crystallographic axes which intersect at 120° and a fourth which is perpendicular to the other three. This fourth axis is usually depicted vertically.

The hexagonal crystal system is divided into the hexagonal and rhombohedral or trigonal divisions. All crystals of the hexagonal division possess a single 6-fold axis of rotation. In addition to the single 6-fold axis of rotation, crystals of the hexagonal division may possess up to six 2-fold axes of rotation. They may demonstrate a centre of inversion symmetry and up to seven mirror planes. Crystals of the trigonal division all possess a single 3-fold axis of rotation rather than the 6-fold axis of the hexagonal division. Crystals of this division may possess up to three 2-fold axes of rotation and may demonstrate a centre of inversion and up to three mirror planes.

Minerals species which crystallise in the hexagonal division are apatite, beryl, and high quartz. Minerals of this division tend to produce hexagonal prisms and pyramids. Example species which crystallise in the rhombohedral division are calcite, dolomite, low quartz, and tourmaline. Such minerals tend to produce rhombohedra and triangular prisms.

Tetragonal

Minerals of the tetragonal crystal system are referred to three mutually perpendicular axes. The two horizontal axes are of equal length, while the vertical axis is of different length and may be either shorter or longer than the other two. Minerals of this system all possess a single 4-fold symmetry axis. They may possess up to four 2-fold axes of rotation, a centre of inversion, and up to five mirror planes.

Mineral species which crystallise in the tetragonal crystal system are zircon and cassiterite. These minerals tend to produce short crystals of prismatic habit.

Orthorhombic

Minerals of the orthorhombic crystal system are referred to three mutually perpendicular axes, each of which is of a different length than the others.

Crystals of this system uniformly possess three 2-fold rotation axes and/or three mirror planes. The holomorphic class demonstrates three 2-fold symmetry axes and three mirror planes as well as a centre of inversion. Other classes may demonstrate three 2-fold axes of rotation or one 2-fold rotation axis and two mirror planes. Species which belong to the orthorhombic system are olivine and barite. Crystals of this system tend to be of prismatic, tabular, or acicular habit.

Monoclinic

Crystals of the monoclinic system are referred to three unequal axes. Two of these axes are inclined towards each other at an oblique angle; these are usually depicted vertically.

The third axis is perpendicular to the other two. The two vertical axes therefore do not intersect one another at right angles, although both are perpendicular to the horizontal axis.

Monoclinic crystals demonstrate a single 2-fold rotation axis and/or a single mirror plane. The holomorphic class possesses the single 2-fold rotation axis, a mirror plane, and a centre of symmetry. Other classes display just the 2-fold rotation axis or just the mirror plane.

Mineral species which adhere to the monoclinic crystal system include pyroxene, amphibole, orthoclase, azurite, and malachite, among many others. The minerals of the monoclinic system tend to produce long prisms.

Triclinic

Crystals of the triclinic system are referred to three unequal axes, all of which intersect at oblique angles. None of the axes are perpendicular to any other axis.

Crystals of the triclinic system may be said to possess only a 1-fold symmetry axis, which is equivalent to possessing no symmetry at all. Crystals of this system possess no mirror planes. The holomorphic class demonstrates a centre of inversion symmetry.

Mineral species of the triclinic class include plagioclase and axinite; these species tend to be of tabular habit.

Crystal Forms

A crystal form is a set of faces which are geometrically equivalent and whose spatial positions are related to one another according to the symmetry of the crystal. If one face of a crystal form is defined, the point symmetry operations which specify the class to which the crystal belongs also determine the other faces of the crystal form.

Fifteen different forms are possible within the isometric or cubic system. These include the hexoctahedron, gyroid, hexatetrahedron, diploid, and tetratoid, among others. The crystal forms of the remaining five crystal systems are the monohedron or pedion, parallelohedron or pinacoid, dihedron, or dome and sphenoid, disphenoid, prism, pyramid, dipyramid, trapezohedron, scalenohedron, rhombohedron, and tetrahedron.

The crystal forms which occur in each crystal class and system must possess a symmetry complementary to that of the associated crystal class and system. For example, a monohedron, which possesses only one face, will never occur in a crystal with inversion symmetry because the inversion operation requires that an equivalent face be present on the opposite side of the crystal.

A simple crystal may consist of only a single crystal form. A more complicated crystal may be a combination of several different forms. All forms which occur in a crystal of a particular system must be compatible with that crystal system.

MONOHEDRON

The monohedral crystal form is also called a pedion. It consists of a single face which is geometrically unique for the crystal and is not repeated by any set of symmetry operations. Members of the triclinic crystal system produce monohedral crystal forms.

PARALLELOHEDRON

The parallelohedral crystal form is also called a pinacoid. It consists of two and only two geometrically equivalent faces which occupy opposite sides of a crystal. The two faces are parallel and are related to one another only by a reflection or an inversion. Members of the triclinic crystal system produce parallelohedral crystal forms.

DIHEDRON

The dihedron consists of two and only two non-parallel geometrically equivalent faces. The two faces may be related by a reflection or by a rotation. The dihedron is termed a dome if the two faces are related only by reflection across a mirror plane. If the two faces are related instead by a 2-fold rotation axis then the dihedron is termed a sphenoid. Members of the monoclinic crystal system produce dihedral crystal forms.

Disphenoid

Members of the orthorhombic and tetragonal crystal systems produce rhombic and tetragonal disphenoids, which possess two sets of non-parallel geometrically equivalent faces, each of which is related by a 2-fold rotation. The faces of the upper sphenoid alternate with the faces of the lower sphenoid in such forms.

Prism

A prism is composed of a set of 3, 4, 6, 8, or 12 geometrically equivalent faces which are all parallel to the same axis. Each of these faces intersects with the two faces adjacent to it to produce a set of parallel edges. The mutually parallel edges of all intersections of the prism sides then form a tube. Prisms are given names based on the shape of their cross-section. Variants of the prism form include the rhombic prism, tetragonal prism, trigonal prism, and hexagonal prism. A prism in which the large faces are divided into two mirror-image faces which intersect with one another at an oblique angle is called a ditetragonal prism, a ditrigonal prism, or a dihexagonal prism. Prisms are associated with the members of the monoclinic crystal system.

Pyramid

A pyramid is composed of a set of 3, 4, 6, 8, or 12 faces which are not parallel but instead intersect at a point. The orthorhombic, tetragonal and hexagonal crystal systems all produce pyramids. These pyramids are named according to the shape of their cross-section in the same way that prisms are. Thus are produced the rhombic pyramid, tetragonal pyramid, trigonal pyramid, and hexagonal pyramid. Each large face of the ditetragonal pyramid, ditrigonal pyramid, and dihexagonal pyramids is divided into two mirror-image faces which occupy an oblique angle with respect to one another.

DIPYRAMID

The dipyramidal crystal form is composed of two pyramids placed base-to-base and related by reflection across a mirror plane which runs parallel to and adjacent to the pyramid bases. The upper and lower pyramids may each have 3, 4, 6, 8, or 12 faces; the dipyramidal form therefore possesses a total of 6, 8, 12, 16, or 24 faces.

The orthorhombic, tetragonal and hexagonal crystal systems all produce dipyramids. These dipyramids are named for the shape of their cross-section just as prisms and pyramids are, resulting in the rhombic dipyramid, trigonal dipyramid, tetragonal dipyramid, and hexagonal dipyramid. The large faces of the ditetragonal, ditrigonal and dihexagonal dipyramids are divided into two mirror-image faces which intersect one another at an oblique angle.

TRAPEZOHEDRON

A trapezohedron is a crystal form possessing 6, 8, or 12 trapezoidal faces. The tetragonal crystal system and both the trigonal and hexagonal divisions of the hexagonal crystal system produce trapezohedral crystal forms. Trigonal trapezohedra possess three trapezoidal faces on the top and three on the bottom for a total of six faces; tetragonal trapezohedra have four faces on top and four on the bottom for a total of eight faces; and hexagonal trapezohedra have six faces on top and six on the bottom, resulting in twelve faces total.

SCALENOHEDRON

A scalenohedron consists of 8 or 12 faces, each of which is a scalene triangle. The faces appear to be grouped into symmetric pairs. The tetragonal and hexagonal crystal systems produce the scalenohedral crystal form, of which examples may be further described as trigonal, tetragonal and hexagonal scalenohedra.

Rhombohedron

The rhombohedral crystal form possesses six rhombus-shaped faces. A rhombohedron resembles in appearance a cube which is poised upright upon one corner and has been either flattened or elongated along an axis which runs diagonally from corner to corner through the centre. The rhombohedral crystal form is produced only by members of the trigonal and rhombohedral divisions of the hexagonal crystal system.

Tetrahedron

A tetrahedron is composed of four triangular faces. In crystals of the isometric system each face is an identical equilateral triangle. In crystals of the tetragonal system each face is an identical isoceles triangle; this variant of the tetrahedron is called a tetragonal tetrahedron. In crystals of the orthorhombic system the faces consist of two pairs of different isoceles triangles; the crystal is then termed a rhombic tetrahedron.

Crystal Classes

The reflection, rotation, inversion, and rotoinversion symmetry operations may be combined in thirty-two different ways. Thirty-two different crystal classes are therefore defined so that each crystal class corresponds to a unique set of symmetry operations. Each of the crystal classes is named according to the variant of a crystal form which it displays. For example, the isometric hexoctahedral class belongs to the isometric crystal system and demonstrates the hexoctahedral crystal form. The rhombic pyramidal, tetragonal pyramidal, *trigonal* pyramidal and hexagonal pyramidal classes each display a variant of the crystal form which is called a pyramid.

Each crystal class is a member of one of the six different crystal systems according to which characteristic symmetry operation it possesses. For example, all crystals of the isometric system possess four 3-fold axes of symmetry, while minerals of the tetragonal system possess a single 4-fold symmetry axis and crystals of the triclinic class show no symmetry at all. The rhombic pyramidal crystal class is thus a member of the orthorhombic crystal system, the tetragonal pyramidal class is a member of the tetragonal crystal system, and the trigonal and hexagonal pyramidal classes are members of the rhombohedral (trigonal) and hexagonal divisions of the hexagonal crystal system respectively.

Table of the 32 Crystal Classes

The following table lists in bold type the six crystal systems. Included are the isometric, hexagonal, tetragonal, orthorhombic, monoclinic, and triclinic systems. The tetragonal crystal system is further separated into the hexagonal and trigonal or rhombohedral divisions. Under each crystal system the table lists by name the crystal classes which occur within that system. For example, the crystal classes which occur within the trigonal crystal system are the trigonal monohedral and trigonal parallelohedral crystal classes. Adjacent to the listing of each crystal class is the symmetry of the class.

When listing the symmetry of each crystal class an axis of rotational symmetry is represented by the capital letter A. Whether this axis is a 2-fold, 3-fold, or 4-fold axis is indicated by a subscript following the letter A. The number of such axes present is indicated by a numeral preceding the capital A. $1A_2$, $2A_3$, and $3A_4$ thus represent one 2-fold axis of rotation, two 3-fold axes, and three 4-fold axes respectively. A centre of inversion is noted by the lowercase letter 'i' while a mirror plane is denoted by 'm'. The numeral preceding the m indicates how many mirror planes are present. Axes of rotary inversion are usually replaced by the equivalent rotations and reflections. For example, a 2-fold rotoinversion axis is equivalent to reflection through a mirror plane perpendicular to the rotoinversion axis. A crystal which possesses a 3-fold rotoinversion axis is equivalent to one which possesses both 3-fold rotational symmetry and inversion symmetry. A 6-fold rotoinversion is equivalent to 3-fold rotation and reflection across a mirror plane at right angles to the rotation axis. The only rotoinversion operation which cannot be thus replaced is 4-fold rotoinversion, which is indicated by R_4.

The class which possesses the highest possible symmetry within each crystal system is termed the holomorphic class of that system. The holomorphic class of each crystal system is indicated in the table by bold type. For example, the triclinic parallelohedron is the holomorphic class of the triclinic crystal system while the isometric hexoctahedron is the holomorphic class of the isomorphic or cubic crystal system. The characteristic symmetry element of each crystal system is listed in bold type. It is thus apparent that the characteristic symmetry element of the isometric crystal system is the possession of four 3-fold axes of rotational symmetry, while the characteristic symmetry element of the rhombohedral division of the hexagonal crystal system is the possession of a single 3-fold axis of rotational symmetry.

Crystal System	*Crystal Class / Crystal Form*	*Symmetry of Class*
Isometric System	hexoctahedron	i, $3A_4$, $4A_3$, $6A_2$,
	gyroid	9m
	hexatetrahedron	$3A_4$, $4A_3$, $6A_2$
	diploid	$3A_2$, $4A_3$, 6m
	tetartoid	i, $3A_2$, $4A_3$, 3m
		$3A_2$, $4A_3$
	dihexagonal dipyramid	i, $1A_6$, $6A_2$, 7m
	hexagonal trapezohedron	$1A_6$, $6A_2$
	dihexagonal pyramid	$1A_6$, 6m
Hexagonal System	ditrigonal dipyramid	$1R_6$, $3A_2$, 3m
Hexagonal Division	hexagonal dipyramid	i, $1A_6$, 1m
	hexagonal pyramid	$1A_6$
	trigonal dipyramid	$1R_6$
	hexagonal	
	scalenohedron	i, $1A_3$, $3A_2$, 3m
Rhombohedral	trigonal trapezohedron	$1A_3$, $3A_2$

Contd...

Crystal System	Crystal Class / Crystal Form	Symmetry of Class
Division	ditrigonal pyramid	$1A_3$, 3m
	rhombohedron	i, $1A_3$
	trigonal pyramid	$1A_3$
	ditetragonal dipyramid	i, $1A_4$, $4A_2$, 5m
Tetragonal System	tetragonal trapzohedron	$1A_4$, $4A_2$
	ditetragonal pyramid	$1A_4$, 4m
	tetragonal scalenohedron	$1R_4$, $2A_2$, 2m
	tetragonal dipyramid	i, $1A_4$, 1m
	tetragonal pyramid	$1A_4$
	tetragonal disphenoid	$1R_4$
	rhombic dipyramid	i, $3A_2$, 3m
Orthorhombic	rhombic disphenoid	$3A_2$
System	rhombic pyramid	$1A_2$, 2m
	prism	i, $1A_2$, 1m
Monoclinic System	sphenoid	$1A_2$
Triclinic System	dome	1m
	parallelohedron	i no symmetry
	monohedron	

Fluid Inclusions

Fluid inclusions are microscopic bubbles of liquid and gas that are trapped within crystals. As minerals often form from a liquid or aqueous medium, tiny blabs of that liquid can become trapped within the crystal structure or in healed fractures within a crystal. These small inclusions range in size from 0.1 to 1 mm and are usually only visible in detail by microscopic study.

These inclusions occur in a wide variety of environments. For example they are found within cementing minerals of sedimentary rocks, in gangue minerals such as quartz or calcite in hydrothermal vein deposits, in fossil amber, and in deep ice cores from the Greenland and Antarctic ice caps. The inclusions can provide information about the conditions existing during the formation of the enclosing mineral.

Hydrothermal ore minerals typically form from high temperature aqueous solutions. The trapped fluid in an inclusion preserves a record of the composition, temperature and pressure of the mineralising environment. An inclusion often contains two or more phases. If a vapour bubble is present in the inclusion along with a liquid phase, simple heating of the inclusion to the point of resorption of the vapour bubble gives a likely temperature of the original fluid. If minute crystals are present in the inclusion, such as halite, sylvite, haematite, or sulphides are present, they provide direct clues as to the composition of the original fluid.

Fluid inclusions are tiny quantities of liquid, vapour, or mixtures of these phases, trapped as impurities within minerals. Their sizes range from submicroscopic up to several hundred micrometres in diameter, and their masses are typically in the order of nanograms to femtograms. Where they are enclosed by minerals that are transparent to visible or infra-red light, fluid inclusions may be observed in a microscope once the host minerals are cut into thin slices and polished. Observations of this kind reveal that rock samples of the Lithosphere commonly contain billions of fluid inclusions. Microscopic study of specific samples permits inclusions that formed during initial mineral growth ("primary inclusions" and "pseudosecondary inclusions") to be discriminated from those that formed some time after mineral growth ("secondary inclusions").

Various sources of evidence suggest that many fluid inclusions preserve the chemical and physical properties of the original parent fluids from which they formed. Fluid inclusions are therefore considered to be direct samples of the volatile phases which circulated through the Lithosphere over the course of the Earth's history, and their chemical analysis provides information on the composition and density of these geologically important phases. Consequently, the results of fluid inclusion research are relevant to our understanding of many natural sub-surface processes in which fluids play a role, such as ore transport and deposition, petroleum generation and migration, explosive volcanism, geothermal energy, earthquake mechanics, petrogenesis of igneous, metamorphic, and diagenetic rocks, contaminant (including radionucleide) transport, and so on.

In addition to exploring applications of fluid inclusion

analyses, current research into fluid inclusions is pursuing several directions, including studies of the systematic behaviour of fluid inclusions in rocks that have undergone long and complicated changes in pressure, temperature and deformation rates; improvements to micro-analytical techniques; experimental synthesis and simulation of the behaviour of natural fluid inclusions; experimental determination of the physicochemical properties of fluids found in natural inclusions; and many more.

Analytical Capabilities

Petrographic Equipment

Petrographic examination of fluid inclusions is one of the most important steps in reconstructing the P-V-T-X evolution of paleo-fluids. The aims of petrography are to discriminate the various compositional types and generations of fluid inclusions in the sample of interest, to identify their mechanism of formation (primary, secondary or pseudosecondary) and their ages relative to each other and relative to the mineral parageneses in which they occur.

Several modes of observation are possible:

- Transparent samples can be observed in visible light
- Inclusions containing aromatic hydrocarbons can be distinguished by their fluorescence in UV light, and their spectral characteristics (intensity versus wavelength of emitted light) can be analysed quantitatively using a microspectrometer
- Inclusions in certain nominally "opaque" minerals can be observed in IR light, and their images made visible using a video camera sensitive to 1100 nm wavelength.

Petrographic images can be recorded as follows:

- Direct digitisation onto a computer disc via the video camera and frame-grabber

- Direct printing on a thermal printer attached to the video or computer monitors
- Photography with conventional SLR cameras mounted on the microscopes

Microthermometry

Next to petrographic examinations, "microthermometry" is the most important analytical technique for characterising fluid inclusions. It involves measuring the temperatures at which phase-transitions are observed to occur in fluid inclusions. If the inclusions have simple compositions (less than 3 or 4 major components) then the microthermometric measurements allow the bulk composition and density of the inclusions to be calculated. For example, the extent to which the melting point of ice is depressed, provides an indication of the bulk salinity of the inclusions. If the inclusions are more complex, then the phase-transition temperatures provide useful constraints on the bulk composition and density, but additional analytical results must be combined to arrive at an exact solution.

Paleoclimate Applications

Trapped bubbles of air and water within fossil amber can be analysed to provide direct evidence of the climate conditions existing when the resin or tree sap formed. The analysis of these trapped bubbles of air provides a record of atmosphere composition going back 140 million years. The data indicate that the oxygen content of the atmosphere reached a high of nearly 35 per cent during the Cretaceous Period and then plummeted to near present levels during the early Tertiary. The abrupt decline corresponds to or closely follows the Cretaceous-Tertiary extinction event and may be the result of a major meteorite impact that created the Chicxulub Crater.

Air bubbles trapped within the deep ice caps can also be analysed for clues to ancient climate conditions.

Fluids and Gases from Tiny Time Capsules

A great many ore-deposit models are tied to the cause of formation of the deposit. Questions about the environmental conditions related to formation of the deposit are temperature, pressure, source of the metals, and composition of any fluids and gases that transported and formed the ore or associated minerals.

Many crystals in the Earth's crust have formed in some kind of fluid. Small quantities of the fluid that surrounded the crystals during growth are commonly trapped as tiny fluid inclusions within these crystals. In many cases, these fluid inclusions are less than 0.1 mm but record important information about the conditions when the ore was being formed.

Trapped in a time capsule the same size as the diameter of a human hair, the ore-forming liquid in this inclusion was so hot and contained so much dissolved solids that when it cooled, crystals of halite, sylvite, gypsum, and haematite formed. As the samples cooled, the fluid shrank more than the surrounding mineral, and created a vapour bubble. Heating the inclusion to the temperature at which the bubble is reabsorbed and daughter crystals dissolve gives an estimate of the minimum temperature at the moment of ore formation.

Current understanding of movements within continents reveals that throughout the Earth's history periods of large-scale fluid movements occurred in the Earth s crust. Some of these fluid migrations resulted in the deposition of metallic ore deposits and accumulations of oil and gas.

Characteristics of fluid inclusions are extremely variable. In the simplest case, when fluid inclusions cool from the elevated temperature at which they formed, the liquid shrinks and separates into a liquid and a vapour bubble. Detailed microthermometric studies give a reasonable estimate of the temperature at which the mineral was formed. Studies of this type reveal that the inclusions were trapped at temperatures from less than 50 C to over 600 C and at pressures equivalent to what is experienced at the Earth s surface and ranging to what would be found several kilometres deep.

Because of the extremely small size of so many fluid inclusions, determining the composition of the trapped fluids is difficult. First, the total amount of dissolved solids is determined by observing with a microscope the freezing/ melting points of the inclusions. The sample is then crushed and rinsed with water. This water is recovered and analysed by using a sensitive analytical technique to determine the ratios of the elements contributed by the trapped fluid. These ratios are used to calculate the composition of the fluid. The compositions range from aqueous solutions with salt content similar to rainwater to fluids with dissolved solid concentrations of over 60 per cent nearly 20 times the amount found in seawater.

Analytical data on fluid inclusions are needed to understand the chemical and physical processes involved in the formation of economic mineral deposits. These data are also critical in understanding modern mineral-deposit models, which promote cost-effective mineral exploration vital to our healthy industrial economy.

6

Single Crystal Spectroscopy

Minerals are elements or combinations of elements in a crystalline form which have been produced by a natural geological process; as such each different mineral species must be different in composition or crystal structure. Therefore, minor substitutions in a mineral structure are not classified as different minerals unless this substitution occurs in constant specific ratios. Minerals have been important since ancient times, due to their wealth creation potential and raw material production.

Minerals have been used as sources of metals, building materials, additives, etc. Some minerals were even used as colorants for medieval manuscripts and cosmetics in ancient Egypt. Identification of minerals is of the utmost importance as knowledge of secondary mineralogy gives a direct indication of the primary mineralogy which is of particular interest for mineral exploration. Identification of minor phases - present as associations - can also give an indication of the geological history of the major mineral species. Therefore, an identification procedure is required to identify these minerals that are non-destructive, fast, and effective.

Current mineral identification techniques include:

1. *X-ray Diffraction (XRD):*

Currently used as the main phase identification technique. However, it is occasionally inappropriate due to the necessity to grind the samples.

2. *Optical Mineralogy:*

Used as an identification procedure but 1 μ m thin sections must be prepared. The procedure is also time consuming with user experience playing a major role in the analysis.

3. *SEM, AA/AE, ICP-MS, XRF:*

The main suite of techniques used for compositional analysis. Carbon coating (SEM) or dissolution (AA/AE, ICP-MS, XRF) is necessary to determine the composition. A set of accurately known standards have to be prepared for a reliable calibration of the method.

Vibrational spectroscopy provides a tool for the identification of minerals with mineral IR spectra being routinely acquired. Identification is easy through a finger printing technique, but the interpretation of spectra is difficult. Vibrational spectroscopy is often the best method for the identification of minerals when other techniques such as XRD can not be undertaken due to sample quantity (minor phases), crystallinity (amorphous/semi-amorphous) and destructive ness (grinding/dissolution). Raman spectroscopy has proven a useful technique for the non-destructive identification of minerals. Copper arsenates, phosphates chlorides, and carbonates have been extensively used in many medieval manuscripts for pigments, but due to their nature, are easily confused owing to their similarities in chemical composition, crystal structure, and colour. The dark green copper phosphate polymorphs - pseudomalachite, reichenbachite, and ludjibaite - are easily confused. These are also confused with malachite

$Cu_2(CO_3)(OH)_2$, and the copper arsenates cornwallite, olivenite, and clinoclase. Easy non-destructive methods of analysis are required for these minerals, thus the utilisation of vibrational spectroscopy is important for the study of these minerals and medieval manuscripts. Raman spectroscopy can easily distinguish different anions such as carbonate, arsenate, phosphate, nitrates, sulphides, chlorides, etc. Minerals with polymorphs can also be distinguished through the use of factor group analysis. Raman spectroscopy has even been employed to identify the geographical origin of minerals. Raman microscopy allows the easy selective orientation of crystals and has the ability to acquire the spectrum of a single crystal due to its spatial resolution of 1μm.

Fluorescence from the sample matrix is also reduced with Raman microscopy due to the reduction of sampling volume. Birefringence can also cause problems in Raman spectroscopy, because of the confocal nature of microscopes; Raman microscopy evades this problem. However, Raman microscopy does suffer from some disadvantages, for instance a large amount of leakage due to the microscope optics can occur. This is due to the analysed and excitation radiation collected without a strictly parallel arrangement. The amount of radiation that is not incident/ collected strictly parallel may be calculated. Samples have an increased tendency to burn/decompose in Raman microscopy with the laser power concentrated into a smaller volume. This is especially a problem with blue samples when using He-Ne lasers as blue samples absorb red light, thus the laser radiation is absorbed by the sample and dissipated as heat.

Solid solutions provide an added hindrance to vibrational spectroscopy due to the broadening of spectra from random substitutions into the crystalline lattice causing a loss of factor group. Standards are often not available for comparison and therefore, identification is sometimes dubious and requiring the synthesis of minerals for reliable identification standards.

Complex solid solutions also exist, the presence of which may be studied by vibrational spectroscopy, as other techniques such as XRD will not detect them. Many secondary mineral samples are also rare and found only in exclusive secondary mineral collections. Techniques such as attenuated total reflectance (ATR), diffuse reflectance (DRIFT) and Raman microanalysis are often used for the study of minerals and other crystalline or semi-crystalline materials. Data analysis is often fraught with danger and a number of critical questions arise. What bands are applicable to which molecular species? How many bands do we expect for each molecule? And how do differences in crystal space group affect the vibrational spectra of the mineral? For these questions to be answered, factor group analysis must be employed. Factor group analysis of simple free molecules is often easy, but the extension into crystalline materials is frequently troublesome.

Factor Group Analysis

Factor group analysis is a method used for determining symmetry of vibrations. It is useful for determining how many bands are expected of each type of molecule in the vibrating unit and is also useful for ascertaining differences in spectra of polymorphous materials. The variation in the spectra of polymorphs are due to the difference in the symmetry of the polymorph. The symmetry elements and the combination of symmetry elements is the crux of factor group analysis. An atom in a vibrating unit, a molecule or crystal, may be thought of as having three degrees of movement, namely x, y, and z directions. From the combination of the symmetry elements of each vibrating unit, the movement of the atoms which result in an overall change in the dipole moment (infrared spectroscopy) or polarisability of the electron cloud (Raman spectroscopy) may be inferred.

Elements of Symmetry

To gain an understanding of factor group analysis it is necessary to first obtain an understanding of elements of symmetry. It is these elements of symmetry and the addition of them that predicts the normal modes of vibration of a molecule in a crystalline lattice. Symmetry elements such as inversion centres, mirror planes, rotation axis and the addition of these to get the overall symmetry species will be looked at later. A symmetry element is an operation that once applied to a shape which leaves that shape unchanged after the operation. One of the simplest symmetry elements is possibly the mirror plane, an example of which is the human hands. If both hands are placed together with small fingers touching it is easy to see that if a mirror is placed between the hands, one hand reflects onto the other hand. Each digit on the hand has an identical image on both the left and right of the image. This is also applicable to molecules such as the water molecule; a mirror plane may be drawn down the centre of the molecule dissecting the oxygen and the angle of the two hydrogens. Another mirror plane, which dissects all atoms is also present.

Another symmetry element is the rotation axis. Other more complex images such as the tetrahedron also have rotation axis. Each apex of the tetrahedron has a three fold rotation axis. If a shape has no symmetry operation applicable to it the symmetry element assigned to a C1 rotation axis and represents the lowest symmetry possible. Inversion centres are centres of symmetry where all elements are able to be transposed through the centre of symmetry to achieve the same image as before. An example of a inversion centre is shown below:

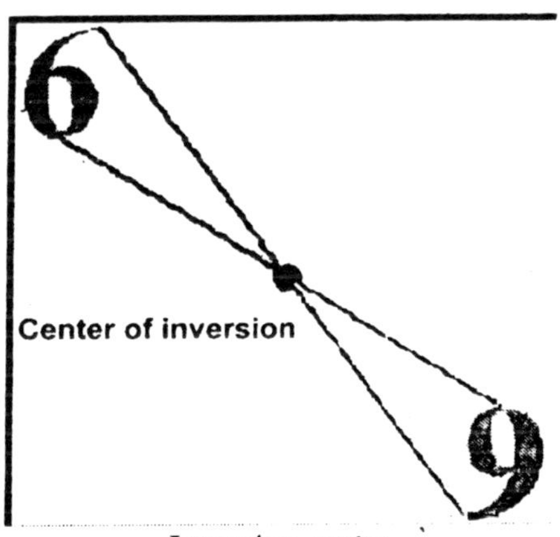

Inversion centre

It can be seen that if all points are transposed through the origin to the other side along the axis, the same shape is achieved. Element of symmetry are able to be combined to achieve a set of possible symmetry operations of a factor group. This factor group will be discussed further in the following sections.

Prediction of the Number of Normal Modes of Vibration

The number of normal modes of vibration for a crystal or molecule may be calculated from the number of degrees of freedom of it has to vibrate. This is related to the number of atoms in the crystal or molecule. Since there are three degrees of freedom for an atom to vibrate, there are 3N modes of motion possible where N is the number of atoms in the molecule or crystal. As the whole crystal or molecule may translate in three directions, then these must be subtracted from the total. Molecules may also rotate in three directions for bent molecules, or two directions for linear molecule, thus three or two modes must be deducted respectively. This allows the following formula for the calculation of the normal modes of vibration of a molecule:

3N-5 Linear molecules

3N-6 Bend molecules.

Crystals have an added confusion, as the contents of the unit cell may not reflect the content of the Bravais unit cell. Only the number of atoms in the Bravais unit cell may be used to predict the number of normal modes of vibration of a crystal. The formula:

$$N = \frac{Z}{LP}$$

This formula may be used to calculate the number of atoms in the Bravais unit cell, where N is the number of atoms in the Bravais unit cell, Z is the number of atoms in the unit cell, and

LP is the number of lattice point as determined from the Table below.

Table: Unit cell type and number of lattice points

Type	A,B,C or I	F	P	R
LP	2	4	1	3 or 1*

* This lattice may have been reduced to primitive form

From the number of atoms in the Bravais cell the number of modes of vibrations of a crystal may be determined by the formula:

3N-3 where N is the number of atoms in the Bravais unit cell.

Three modes are subtracted in the case of a crystal as the rotation of a atom or a molecule is not the same as the rotation of a neighbouring group or atom.

Point Group Analysis of Free Molecules

The water molecule is one of the simplest cases which contains a combination of symmetry elements. The water molecule has three atoms - one oxygen bound to two hydrogens – at an angle of ~120°. A line can be drawn down the middle of the oxygen, dissecting the angle between the two hydrogens. If the molecule is rotated around this axis, there are two positions, that produce the same image as the original starting position. This line represents a two fold rotation or commonly referred to as the C2 axis and a rotation of 180° about the C2 axis is called a C2 symmetry operation.

There are also two more symmetry elements in the water molecule, the mirror planes. The first plane is the plane of the atoms, the second plane is normal to the first and intersects it along the C2 axis. These mirror planes reflect the atoms on one side of the molecule directly onto the other side. These are the Cs planes, where s represents a mirror plane. Once the symmetry operations for the molecule are known, the point group may be determined by following the flow chart. In this case, it may be deduced that the molecule has C2v symmetry.

The next step is to use the point group to assign the symmetry species of the vibrations of the molecule. Consider the water molecule in terms of distinct elements; the oxygen, which is sitting on a C2 axis and two mirror planes, and two identical hydrogens sitting on a mirror plane. There are rules of convention governing the assignment of the x, y, z axes. The z axis in vibrational spectroscopy is considered the most important axis, which is commonly the axis that intercepts the most atoms with the most symmetry. In the water molecule the z axis corresponds to the C2 axis. In this case the oxygen atom is sited on the highest point of symmetry. The x axis is now chosen to be perpendicular to the molecular plane and the y axis to be in the molecular plane.

Having assigned the axes, correlation tables such as those in Farmer may be consulted, to find the symmetry species of the vibrations of the molecules. The first column shows the highest symmetry group. In this particular case it is the C2v column that is the highest symmetry species present. The letters in this column represent the modes, which are available to the molecule to vibrate. Each distinct atom (with respect to symmetry, and atomic type) is considered for its particular symmetry elements. Consulting the correlation for oxygen (C2v), it can be seen that there is displacement for the oxygen in the A1, B1 and B2 symmetry operations.

The hydrogens can be seen to be on C_5^{yz} elements allowing for two displacements in the A1 mode - one in the A2 mode, one for B1 and two for the B2 mode. The addition of these modes gives the reducible representation given by in this case 3A1 + A2 + 2B1 + 3B2. However six of these modes describe the modes of vibration which describe rotations and translations. The available modes of vibration of the water molecules can then be obtained by the subtraction of the translation and rotations possible.

Table: Correlation table for the C2v point group

C2v- mmm2	*1* *C2v*	*2* *C2*	*2* *Czxs*	*2* *Cyzs*	*4* *C1*
xx,yy,zz, Tz	A1 1z	A 1z	A′ 2p	A′ 2p	A 3
xy, Rz	A2 0	A 1z	A″ 1y	A″ 1x	A 3
zx, Tx, Ry	B1 1x	B 2p	A′ 2p	A″ 1x	A 3
yz, Ty, Rx	B2 1y	B 2p	A″ 1y	A′ 2p	A 3

Czxs: 1ef;11d;14c;15c;18d; 20c.

Cyzs: 1gh;2ab;4c;7a;11e;12a;14de;16b;16b;18c;20d;22b.

Table: Factor group analysis of the water molecule

	Elements of symmetry	*Available modes* *A1*	*A2*	*B1*	*B2*
Displacements of the O atom	C2v	1z	0	1x	1y
Displacements of the H atoms	Csyz	2p	1x	1x	2p
Rotations and translations		Tz	Rz	Tx, Ry	Ty, Rx
Available vibrations of the		2A1	-	-	1B2

The irreducible representation of the water molecule is G = 2A1 + B2, which predicts 3 modes for water. The 3N-6 rule discussed earlier allows for a check of the number of modes possible, in this case confirming the prediction of three modes of vibration for water. It is seen that the translation apply to the normal modes for the water molecule. This indicates that the Diploe moment oscillates in the same direction as the translation allowing for infrared activity. It also can be seen that the polarisable tensor also has modes applicable for the normal modes of water thus the modes will be Raman active also.

Application of the internal modes to the crystal structure

The discussion to date has deal with the vibrational modes available to isolated molecules. For the extension of factor group analysis to crystalline solids an understanding of crystallography is required. Crystalline lattices exist is a specific structure for which the atoms may only exist on only specific sites. There exist some 230 space groups which relate to 33 crystallographic space groups. Each specific space group has a finite number of sites with specific symmetry. The "International Table For Crystallography" has the relative coordinates for each symmetry site contained in the unit cell which may be used in conjunction with single crystal XRD to determine the identity of the atoms on each site.

Once the site point group has been determined it is necessary to check that there is no choice between the similar symmetry species. For instance, the cases of the water molecule were a choice of the Cs mirror plane in the yx or xz directions. In this case the Wyckoff letter in combination with the Schoenflies superscript will yield the answer.

Once the sites have unequivocally been determined, the factor group splitting can be determined. Appropriate correlation tables can be consulted to determine how the site symmetry reduces the internal modes to lower symmetry modes. If vibrational modes are degenerate such as E and T modes, the site symmetry may split the modes into lower symmetry species allowing multiple modes for the degenerate modes. The crystal symmetry correlation table, (Factor Group) can be used to determine the splitting of the site mode will result in splitting of these modes.

The occurrence of more than one of the identical molecules (with relation to symmetry and atom type, could result in coupling. In the case of water molecules (C2v) sitting on a C1 site in a C2h crystal, such as the gypsum case, the A1 and B2 mode form A modes on C1 site and split to Ag, Bg, Au, and Bu modes in the C2h crystal. This would suggest that in this case there is the occurrence of four of the same molecules in the lattice, which will result in the splitting of A modes. It is also noted that the results in a symmetric characteristic (Ag modes, vibrating in phase) and an anti-symmetric mode (Bg, vibrating out of phase).

Table: Factor group splitting for a water molecule on a C1 site in a C2h crystal

Free *C2v*	*Site* *C1*	*Crystal* *C2h*
2A1	3A	3Ag
B2		3Bg
		3Au
		3Bu

The 3N-3 rule may then be applied to determine if all modes of vibration have been described. In the case of crystals, the contents of the Brauvis lattice must be considered and not the total unit cell contents. Taking this into account, three modes are available to the free water molecule, but in the case above there are 4 water molecules in the Brauvis cell and hence each mode splits into 4, allowing 12 modes of vibration. All the modes of vibration determined in this section have been concerned with the vibration of the molecule, the so-called internal modes of vibration. There are also the vibrations of the molecule as a whole, vibrating within the crystalline lattice.

Lattice Modes

Lattice modes are dealt with a similar mechanism to that of the vibrations of molecules. The main difference between the factor group analysis of lattice vibrations and internal modes, is that lattice modes are considered as the movements of groups of atoms in the crystalline lattice. For instance, gypsum $CaSO_4.2H_2O$, the water and sulphate molecules would be considered as free entities then subjected to the symmetry constraints of the site, then the crystal symmetry. It is then necessary to consider the motions of the sulphate group, the water molecules, as well as the motion of the calcium atoms in the crystal. To consider the factor group analysis of lattice modes it is necessary to know the site symmetry of each group. It is necessary to know the modes available for the molecule to translate and rotate. While in the case of atoms only translations need be considered due to their rotational symmetry and molecules considered to translate and rotate. The appropriate modes may be deduced from correlation tables. In the case of water molecules on C1 sites there is 6A modes available for the water molecule to rotate and translate. This is then related to the C2h crystal group. This predicts the formation of the reducible representation 6Ag + 6Bg + 6Au + 6Bu. Each group of atoms is considered in turn to produce the active lattice vibrations which are added to achieve the lattice modes of the crystal. It is again necessary to consider the three modes that represent the translations of the whole crystal, which are not active vibrations. These modes are subtracted to give the irreducible representation 5Ag + 4Bg + 5Au + 4Bu.

Assignment of spectra, Selection Rules, Polarised Orientated Vibrational spectroscopy

Assignment of spectral bands to various symmetry species can be achieved via polarised single crystal Raman and polarised single crystal infrared investigations. The symmetry species of the vibrations can be ascertained by orienting the crystal of a mineral along different polarisations of the incident radiation. For Raman measurements the polarisability tensor of the vibration is different along different axis of the crystal. The probability of a Raman transition is proportional to:

$\int \psi o \alpha \psi i \, d\tau$ Where ψo is the vibrational wave function, α is the polarisable of the molecule, ψi is the vibrational wave function, and $d\tau$ implies the integration is carried over all possible variables of the wave functions. The polarisable α is the ease at which the electron cloud in the molecule may be distorted. This quantity is a tensor, a 3*3 array of components, i.e. α xx, α xy, α xz, α yx, α yy, α yz, α zx, α zy, and α zz, where α kl = α lk. In the case where the integral is zero, then there is zero probability that the transition will occur. The integral will only be non- zero in the case where the product $\psi o \alpha \psi$ is totally symmetric, that is +1. This can only be achieved when α kl has the same symmetry properties as ψi. The symmetry properties of αkl is the same as the symmetry properties of kl, thus if the normal mode has the same symmetry as one of these binary combinations of x, y, and z, then a transition from the ground state will be Raman active. If a character is shown, then the transition along the axis may be allowed. Modes of vibration that are not allowed will be of zero intensity or very low intensity. Orientated single crystal Raman studies, with Raman spectra taken of all orientations and polarisations possible, will yield the symmetry species of the vibrations.

The polarisability tensors for a vibration are calculated along spectroscopic axis. Thus it is necessary to know the relationship of the crystallographic axis in relation to the spectroscopic axis. This can be achieved by knowledge of the translations of the spectroscopic group, available from correlation tables. In this way a spectroscopic axis translation can be seen to move in a direction associated with a mode of vibration. From the correlation tables these modes can be related to the site and space groups to determine the spectroscopic axis with relation to the crystal axis. From an atomic coordinate model, the relationship between the spectroscopic and crystallographic axis can be deduced. The spectroscopic axis can also be deduced from the correlation tables, by looking at the symmetry species and Schoenflies / Wyckoff letter if the space group has choices for the definition of the same symmetry element.

Once the symmetry species of the bands are identified there is a possibility of further assignment from deuteration, isotopic and analogous series studies (cationic and anionic series) through the use of Hooke's Law. Satellite bands are known to exist in cerussite, with the intensity of these bands about 1 per cent of the corresponding n1 band intensity. This satellite is due to isotopic 180 substitution in the carbonate group. The calculated frequency of this vibration from Equation

1 for cerussite 1030, aragonite 1058, witherite 1034, and strontianite 1044 cm-1 which are in good agreement with the experimental results.

$$V = \frac{1}{2\pi c}\sqrt{\frac{k_o}{\mu}}$$

μ = reduced mass of participating atoms

ko = force constant

V = vibrational frequency in wavenumbers

c= speed of light

Studies have shown a strong correlation between OH stretching frequencies and both the O...O bond distances and the H...O hydrogen bond distances. The significant work of Libowitzky showed that a regression function can be employed, relating the above correlations with regression coefficients better than 0.96. Minerals where two or more types of OH units are identified in the structure, the known hydrogen bond distances may be used to calculate the hydroxyl stretching frequency. From this predicted hydroxyl stretching frequency further assignment of the spectra may be made.

Some vibrations that are treated as lattice vibrations such as hydroxyl stretching vibrations therefore may be hard to assign due to there consideration with the other lattice modes. Vector analysis may be employed to gain insight into these bands. Vector analysis considers vibrations of a molecule as a vector. Interactions of this vector with other vectors of an identical nature is deduce from the factor group. For the below example of olivenite, it is unclear how many vibrations are expected from the hydroxyl groups.

As the hydroxyl groups are siting on C1 sites there is only one vector that the hydroxyl is allowed to vibrate on their site symmetry (A). When this is allowed to split in the factor group there is 1Ag, 1Au, 1Bg, and 1Bu modes predicted. Of which, g modes are Raman active and u modes are infrared active. Olivenite does indeed have two hydroxyl bands in the Raman.

This implies that in the structure of olivenite there are four hydroxyls which couple to produce the four modes.

Birefringence

Birefringence is a well known hindrance to single vibrational studies. Birefringence is the differences in the refractive indices down each independent optical axis. As such, there is a difference in the velocity of light down these axis, hence a change in the resulting polarisation of the incident light. Erythrite is known to be birefringent with refractive indicis of $\alpha = 1.625$, $\beta = 1.661$, and $\gamma = 1.691$. The optical axis's or pseudo-optical axis's are orthogonal on all axis except on the 010 face which the optical axis make a 30° angle with the c crystal axis. The amount of depolarisation of incident radiation on a birefringent crystal can be calculated from the equation:

$$L = 100[\cos 2\Phi - \sin 2(\Gamma - \Phi).\sin 2\Gamma.\sin 2(\Delta / \lambda).180]$$

where L is the amount of radiation for which the plane of polarisation is rotated by 90°, Γ is the angle the optical axis makes with the incident radiation, Φ is the angle between the analyser and polariser, Δ is the difference between the refractive indicis multiplied by the thickness of the sample and λ in the wavelength of the incident radiation. This equation predicts along the 010 face of a crystal of erythrite incident polarised radiation is 28 per cent scrambled. This prevents the collection of reliable oriented single crystal data from the 010-crystal face.

Porto Notation

Porto notation is a way of expressing the orientation of the crystal with reference to the polarisation of the laser in both the excitation and analysing directions. This notation consists of a four letters code expressing the direction of the propagation of the incident radiation, the polarisation of the incident radiation, the polarisation of the analysed radiation and the direction of propagation of the analysed radiation. For example

CABC, the first letter depicts the axis for the propagation of the laser light, the second depicts the orientation of the polarisation direction, i.e. the direction that the laser is vibrating in the excitation, the third is the vibrational direct of the Raman scattered photons that are being analysed and the last the direction of propagation of the analysed laser beam. This is all done with reference to the crystal that is being analysed.

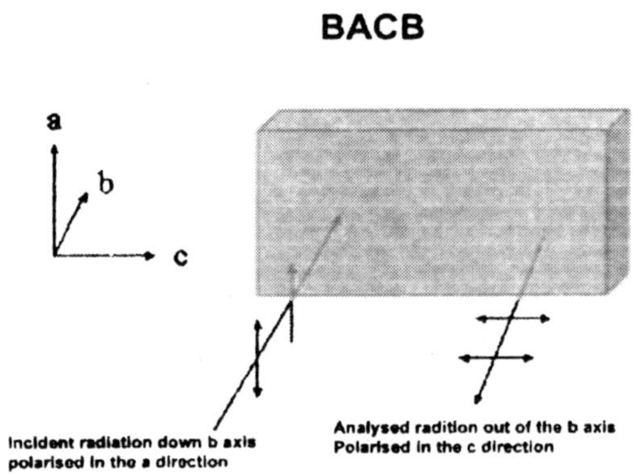

An example of Porto notation.

LO/TO Modes

Optical phonons propagate transversely or longitudinally in the crystal. If there is a change in the dipole moment involved in the vibrational mode, the frequency of the phonon may vary between situations where the dipole moment is aligned either transversely or longitudinally to the direction of propagation.

Long-range interaction between the dipole and the polarisation field decreases the frequency of the vibration of the phonon from that of the isolated vibration. The magnitude of the splitting depends on the magnitude of the change in the dipole moment of the vibration (i.e. the intensity of the IR band).

For C46n space group, the A1 mode is polarised in the Z direction. Using the Raman microscope (i.e. 180° scattering), the phonon detected will be parallel to the axis of the laser. So for a Z(XX)Z or Z(YY)Z experiment, the phonon wave vector is in the Z direction, which is parallel to the change in the dipole moment.

Thus, the phonon travels longitudinally wrt to the change in dipole moment of the vibration and will be LO in character. On the other hand, Y(XX)Y will be TO as the phonon is along the Y axis and is transverse to the direction of polarisation of the vibration (z).

In situations other than 180° scattering, draw up a wave vector diagram on the basis of conservation of momentum. For a C6n factor group, the yz active E1 mode has a change in the dipole moment along the y axis (Y polarisation). Therefore the E1 mode will be split into LO (y component) and TO (z component) bands.

In summary:

(1) For splitting to occur the mode must be IR active – Raman active g modes will not exhibit splitting.

(2) The size of the splitting depends on the IR intensity of the band.

(3) Determine the direction of travel of the phonon.

(4) Look up the polarisation of the mode in the appropriate character tables. Note that Turrell gives the polarisation of the particular modes when there is degeneracy, e.g. E1(-X), E1(Y). The mode can be assigned to TO or LO if the polarisation is transverse or longitudinal to the direction of the phonon, respectively. When the direction of the phonon can be resolved into a polarised and non-polarised axis, the mode will show both LO and TO components.

(5) The magnitude of the splitting is angular dependent and can be calculated (apparently).

7

Skarns and Skarn Deposits

Although there are earlier descriptions of deposits now known to contain skarn, the first published use of the term "skarn" was by Tornebohm.

For other pre-1970 publications concerning skarn deposits, Burt (1982) provides a very useful annotated historical bibliography. Einaudi et al. (1981), a fairly exhaustive review paper on skarn deposits, is a good source of other references and definitions, some of which are summarised below. Another source of historical information is the recently published field trip guide to the classic Banat region of Romania.

Definitions: There are many definitions and usages of the word "skarn". Skarns can form during regional or contact metamorphism and from a variety of metasomatic processes involving fluids of magmatic, metamorphic, meteoric, and/or marine origin. They are found adjacent to plutons, along faults and major shear zones, in shallow geothermal systems, on the bottom of the seafloor, and at lower crustal depths in deeply buried metamorphic terrains. What links these diverse environments, and what defines a rock as skarn, is the mineralogy. This mineralogy includes a wide variety of calc-silicate and associated minerals but usually is dominated by garnet and pyroxene.

Skarns can be subdivided according to several criteria. Exoskarn and endoskarn are common terms used to indicate a sedimentary or igneous protolith, respectively. Magnesian and calcic skarn can be used to describe the dominant composition of the protolith and resulting skarn minerals. Such terms can be combined, as in the case of a magnesian exoskarn which contains forsterite-diopside skarn formed from dolostone.

Calc-silicate hornfels is a descriptive term often used for the relatively fine-grained calc-silicate rocks that result from metamorphism of impure carbonate units such as silty limestone or calcareous shale. Reaction skarns can form from isochemical metamorphism of thinly interlayered shale and carbonate units where metasomatic transfer of components between adjacent lithologies may occur on a small scale (perhaps centimetres) (e.g. Vidale, 1969; Zarayskiy et al., 1987). Skarnoid is a descriptive term for calc-silicate rocks which are relatively fine-grained, iron-poor, and which reflect, at least in part, the compositional control of the protolith (Korzkinskii, 1948; Zharikov, 1970). Genetically, skarnoid is intermediate between a purely metamorphic hornfels and a purely metasomatic, coarse-grained skarn.

For all of the preceding terms, the composition and texture of the protolith tend to control the composition and texture of the resulting skarn. In contrast, most economically important skarn deposits result from large scale metasomatic transfer, where fluid composition controls the resulting skarn and ore mineralogy. This is the mental image that most people share of a "classic" skarn deposit. Ironically, in the "classic" skarn locality described by Tornebohm at Persberg, skarn has developed during regional metamorphism of a mostly calcareous Proterozoic iron formation. This reinforces the importance of Einaudi et al.'s (1981) warning that the words "skarn" and "skarn deposits" be used strictly in a descriptive sense, based upon documented mineralogy, and free of genetic interpretations.

Not all skarns have economic mineralisation; skarns which contain ore are called skarn deposits. In most large skarn deposits, skarn and ore minerals result from the same hydrothermal system even though there may be significant differences in the time/space distribution of these minerals on a local scale. Although rare, it is also possible to form skarn by metamorphism of pre-existing ore deposits as has been suggested for Aguilar, Argentina (Gemmell et al., 1992), Franklin Furnace, USA (Johnson et al., 1990), and Broken Hill, Australia (Hodgson, 1975).

Skarn Mineralogy

Just as mineralogy is the key to recognising and defining skarns, it is also critical in understanding their origin and in distinguishing economically important deposits from interesting but uneconomic mineral localities. Skarn mineralogy is mappable in the field and serves as the broader "alteration envelope" around a potential ore body. Because most skarn deposits are zoned, recognition of distal alteration features can be critically important in the early exploration stages. Details of skarn mineralogy and zonation can be used to construct deposit-specific exploration models as well as more general models useful in developing grass roots exploration programmes or regional syntheses.

Although many skarn minerals are typical rock-forming minerals, some are less abundant and most have compositional variations which can yield significant information about the environment of formation (e.g. pyroxene - Takano, 1998; scapolite - Pan, 1998).

Some minerals, such as quartz and calcite, are present in

almost all skarns. Other minerals, such as humite, periclase, phlogopite, talc, serpentine, and brucite are typical of magnesian skarns but are absent from most other skarn types. Additionally, there are many tin, boron, beryllium, and fluorine-bearing minerals which have very restricted, but locally important, parageneses.

The advent of modern analytical techniques, particularly the electron microprobe, makes it relatively easy to determine accurate mineral compositions and consequently, to use precise mineralogical names. However, mineralogical names should be used correctly so as not to imply more than is known about the mineral composition. For example, the sequence pyroxene, clinopyroxene, calcic clinopyroxene, diopsidic pyroxene, and diopside, are increasingly more specific terms. Unfortunately, it is all too common in the geologic literature for specific end member terms, such as diopside, to be used when all that is known about the mineral in question is that it might be pyroxene.

Zharikov (1970) was perhaps the first to describe systematic variations in skarn mineralogy among the major skarn classes. He used phase equilibria, mineral compatibilities, and compositional variations in solid solution series to describe and predict characteristic mineral assemblages for different skarn types. His observations have been extended by Burt (1972) and Einaudi et al. (1981) to include a wide variety of deposit types and the mineralogical variations between types. The minerals which are most useful for both classification and exploration are those, such as garnet, pyroxene, and amphibole, which are present in all skarn types and which show marked compositional variability. For example, the manganiferous pyroxene, johannsenite, is found almost exclusively in zinc skarns. Its presence, without much further supporting information, is definitive of this skarn type.

When compositional information is available, it is possible to denote a mineral's composition in terms of mole per cent of the end members. For example, a pyroxene which contains

70 mole per cent hedenbergite, 28 mole per cent diopside, and 2 mole per cent johannsenite could be referred to as Hd70Di28Jo2. In many skarn systems, variation in iron content is the most important parameter and thus, many minerals are described simply by their iron end member, e.g. Hd10 or Ad90. Large amounts of compositional information can be summarised graphically. Triangular plots commonly are used to express variations in compositionally complex minerals such as garnet and pyroxene.

Amphiboles are more difficult to portray graphically because they have structural as well as compositional variations. The main differences between amphiboles in different skarn types are variations in the amount of Fe, Mg, Mn, Ca, Al, Na, and K. Amphiboles from Au, W, and Sn skarns are progressively more aluminous (actinolite-hastingsite-hornblende), amphiboles from Cu, Mo, and Fe skarns are progressively more iron-rich in the tremolite-actinolite series, and amphiboles from zinc skarns are both Mn-rich and Ca-deficient, ranging from actinolite to dannemorite. For a specific skarn deposit or group of skarns, compositional variations in less common mineral phases, such as idocrase, bustamite, or olivine, may provide insight into zonation patterns or regional petrogenesis (e.g. Giere, 1986; Agrell and Charnely, 1987; Silva and Siriwardena, 1988; Benkerrou and Fonteilles, 1989).

Evolution of Skarns in Time and Space

As was recognised by early skarn researchers (e.g. Lindgren 1902; Barrell, 1907; Goldschmidt, 1911; Umpleby, 1913; Knopf, 1918), formation of a skarn deposit is a dynamic process. In most large skarn deposits there is a transition from early/distal metamorphism resulting in hornfels, reaction skarn, and skarnoid, to later/proximal metasomatism resulting in relatively coarse-grained ore-bearing skarn. Due to the strong temperature gradients and large fluid circulation cells caused by intrusion of a magma (Norton, 1982; Salemink and Schuiling, 1987; Bowers et al., 1990), contact metamorphism can be considerably more complex than the simple model of isochemical recrystallisation typically invoked for regional metamorphism.

For example, circulating diverse fluids through a fracture in a relatively simple carbonate protolith can result in several different reactions. Thus, the steep thermal gradients common in most plutonic environments, result in complex metamorphic aureoles complete with small-scale metasomatic transfer as evidenced by reaction skarns and skarnoid.

More complex metasomatic fluids, with the possible addition of magmatic components such as Fe, Si, Cu, etc., produce a continuum between purely metamorphic and purely metasomatic processes. This early metamorphism and continued metasomatism at relatively high temperature (Wallmach and Hatton, 1989, describe temperatures > 1200C) are followed by retrograde alteration as temperatures decline. A link between space and time is a common theme in ore deposits and requires careful interpretation of features.

One of the more fundamental controls on skarn size, geometry, and style of alteration is the depth of formation. Quantitative geobarometric studies typically use mineral equilibria (Anovitz and Essene, 1990), fluid inclusions (Guy et al., 1989) or a combination of such methods (Hames et al., 1989) to estimate the depth of metamorphism. Qualitative methods include stratigraphic or other geologic reconstruc tions and interpretation of igneous textures. Simple observations of chilled margins, porphyry groundmass grain size, pluton morphology, and presence of brecciation and brittle fracture allow field distinctions between relatively shallow and deep environments.

The effect of depth on metamorphism is largely a function of the ambient wall rock temperature prior to, during, and post- intrusion. Assuming an average geothermal gradient for an orogenic zone of about 35C per kilometre (Blackwell et al., 1990), the ambient wall rock temperature prior to intrusion at

2 km would be 70C, whereas at 12 km it would be 420C. Thus, with the added heat flux provided by local igneous activity, the volume of rock affected by temperatures in the 400-700C range would be considerably larger and longer lived surrounding a deeper skarn than a shallower one. In addition, higher ambient temperatures could affect the crystallisation history of a pluton as well as minimise the amount of retrograde alteration of skarn minerals.

At a depth of 12 km with ambient temperatures around 400C, skarn may not cool below garnet and pyroxene stability without subsequent uplift or other tectonic changes. The greater extent and intensity of metamorphism at depth can affect the permeability of host rocks and reduce the amount of carbonate available for reaction with metasomatic fluids. An extreme case is described by Dick and Hodgson (1982) at Cantung, Canada, where the "Swiss cheese limestone" was almost entirely converted to a heterogeneous calc-silicate hornfels during metamorphism prior to skarn formation. The skarn formed from the few remaining patches of limestone has some of the highest known grades of tungsten skarn ore in the world (Mathiason and Clark, 1982).

The depth of skarn formation also will affect the mechanical properties of the host rocks. In a deep skarn environment, rocks will tend to deform in a ductile manner rather than fracture. Intrusive contacts with sedimentary rocks at depth tend to be sub-parallel to bedding; either the pluton intrudes along bedding planes or the sedimentary rocks fold or flow until they are aligned with the intrusive contact.

Examples of skarns for which depth estimates exceed 5-10 km include Pine Creek, California (Brown et al., 1985) and Osgood Mountains, Nevada (Taylor, 1976). In deposits such as these, where intrusive contacts are sub-parallel to bedding planes, skarn is usually confined to a narrow, but vertically extensive, zone. At Pine Creek skarn is typically less than 10 m wide but locally exceeds one kilometre in length and vertical extent (Newberry, 1982).

Thus, skarn formed at greater depths can be seen as a narrow rind of small size relative to the associated pluton and its metamorphic aureole. In contrast, host rocks at shallow depths will tend to deform by fracturing and faulting rather than folding. In most of the 13 relatively shallow skarn deposits reviewed by Einaudi (1982a), intrusive contacts are sharply discordant to bedding and skarn cuts across bedding and massively replaces favourable beds, equalling or exceeding the (exposed) size of the associated pluton. The strong hydrofracturing associated with shallow level intrusions greatly increases the permeability of the host rocks, not only for igneous-related metasomatic fluids, but also for later, possibly cooler, meteoric fluids (Shelton, 1983). The influx of meteoric water and the consequent destruction of skarn minerals during retrograde alteration is one of the distinctive features of skarn formation in a shallow environment.

The shallowest (and youngest) known skarns are presently forming in active geothermal systems (McDowell and Elders, 1980; Cavarretta et al., 1982; Cavarretta and Puxeddu, 1990) and hot spring vents on the seafloor (Zierenberg and Shanks, 1983). These skarns represent the distal expression of magmatic activity and exposed igneous rocks (in drill core) are dominantly thin dikes and sills with chilled margins and a very fine grained to aphanitic groundmass.

The degree to which a particular alteration stage is developed in a specific skarn will depend on the local geologic environment of formation. For example, metamorphism will likely be more extensive and higher grade around a skarn formed at relatively great crustal depths than one formed under shallower conditions. Conversely, retrograde alteration during cooling, and possible interaction with meteroric water, will be more intense in a skarn formed at relatively shallow depths in the earth's crust compared with one formed at greater depths. In the deeper skarns carbonate rocks may deform in a ductile manner rather than through brittle fracture, with bedding parallel to the intrusive contact; in shallower systems the reverse may be true. These differences in structural style will in turn affect the size and morphology of skarn. Thus, host rock composition, depth of formation, and structural setting will all cause variations from the idealised "classic" skarn model.

Au, Cu, Fe, Mo, Sn, W, and Zn-Pb Skarn Deposits

Groupings of skarn deposits can be based on descriptive features such as protolith composition, rock type, and dominant economic metal(s) as well as genetic features such as mechanism of fluid movement, temperature of formation, and extent of magmatic involvement. The general trend of modern authors is to adopt a descriptive skarn classification based upon the dominant economic metals and then to modify individual categories based upon compositional, tectonic, or genetic variations.

This is similar to the classification of porphyry deposits into porphyry copper, porphyry molybdenum, and porphyry tin types; deposits which share many alteration and geochemical features but are, nevertheless, easily distinguish able. Seven major skarn types (Au, Cu, Fe, Mo, Sn, W, and Zn-Pb) have received significant modern study and several others (including F, C, Ba, Pt, U, REE) are locally important. In addition, skarns can be mined for industrial minerals such as garnet and wollastonite.

Major Skarn Types:

- Fe Skarns
- Au Skarns
- W Skarns
- Cu Skarns
- Zn Skarns
- Mo Skarns
- Sn Skarns

IRON SKARNS

The largest skarn deposits are the iron skarns. Major reviews of this deposit type include Sangster (1969), Sokolov and Grigorev (1977), and Einaudi et al. (1981). Iron skarns are mined for their magnetite content and although minor amounts of Cu, Co, Ni, and Au may be present, iron is typically the only commodity recovered (Grigoryev et al., 1990). Many deposits are very large (>500 million tons, >300 million tons contained Fe) and consist dominantly of magnetite with only minor silicate gangue. Some deposits contain significant amounts of copper and are transitional to more typical copper skarns (e.g. Kesler, 1968; Vidal et al., 1990).

Calcic iron skarns in oceanic island arcs are associated with iron-rich plutons intruded into limestone and volcanic wall rocks. In some deposits, the amount of endoskarn may exceed exoskarn. Skarn minerals consist dominantly of garnet and pyroxene with lesser epidote, ilvaite, and actinolite; all are iron-rich (Purtov et al., 1989). Alteration of igneous rocks is common with widespread albite, orthoclase, and scapolite veins and replacements, in addition to endoskarn.

In contrast, magnesian iron skarns are associated with diverse plutons in a variety of tectonic settings; the unifying feature is that they all form from dolomitic wall rocks. In magnesian skarns, the main skarn minerals, such as forsterite, diopside, periclase, talc, and serpentine, do not contain much iron; thus, the available iron in solution tends to form magnetite rather than andradite or hedenbergite (e.g. Hall et al., 1989).

Overprinting of calcic skarn upon magnesian skarn is reported from many Russian deposits (Sokolov and Grigorev, 1977; Aksyuk and Zharikov, 1988). In addition, many other skarn types contain pockets of massive magnetite which may be mined for iron on a local scale (e.g. Fierro area, New Mexico, Hernon and Jones, 1968). Most of these occurrences form from dolomitic strata or from zones that have experienced prior magnesian metasomatism (e.g. Imai and Yamazaki, 1967).

Gold Skarns

Prior to the dramatic rise in the price of gold in the early 1970s, most gold produced from skarn deposits came as a by-product of the mining of other metals, particularly Cu. The one notable exception was the Nickel Plate mine in the Hedley district, British Columbia, which had been mined for high grade gold in skarn from the turn of the century (Billingsley & Hume 1941). This deposit has been intensively studied (Ray *et al.* 1986b, 1988, 1993, 1995, 1996; Ettlinger 1990; Ettlinger *et al.* 1992; Ray & Dawson 1987, 1988, 1994) and has served as a de facto exploration model for gold skarn deposits in combination with the relatively similar Fortitude deposit in Nevada.

Numerous other gold skarn deposits have been discovered in the past several decades which differ in important ways from the Hedley-Fortitude model. For example, some are magnesian skarns (Butte Highlands, Montana, Ettlinger *et al.* 1996; Marvel Loch, Australia, Mueller 1991, Mueller *et al.* 1991), some are magnetite-dominant (Bermejal, Mexico, de la Garza *et al.* 1996; Key East, Washington, Lowe 1998), some are garnet-dominant and relatively oxidised (Ban Na Lom, Thailand, Pisutha-Arnond *et al.* 1984; McCoy, Nevada, Brooks 1994; Nambija, Ecuador, Hammarstrom 1992; Red Dome, Australia, Ewers & Sun 1989; Wabu, Irian Jaya, Allen *et al.* 1998), and some occur in iron-rich rocks in regional metamorphic terrains (Lucky Draw, Australia, Sheppard *et al.* 1995; Lupin, Northwest Territories, Lhotka & Nesbitt 1989; Mallapakonda and Oriental, India, Siddaiah & Rajamani 1989; Navachab, Namibia, Noertemann 1997, Moore and Jacob, 1998; Nevoria, Australia, Mueller 1997; Tillicum, British Columbia, Ray *et al.* 1986a; Peterson 1996). Reviews of gold-bearing skarns that contain useful background data include: Yakrushev (1972), Meinert (1989, 1998), Ray *et al.* (1990), Theodore *et al.* (1991), and Ray & Webster (1991, 1995).

The term "gold skarn" is used here in the economic sense suggested by Einaudi *et al.* (1981) and refers to ore deposits that are mined solely or predominantly for gold and which exhibit calc-silicate alteration, usually dominated by garnet and pyroxene, that is related to mineralisation. This usage excludes deposits such as Big Gossan (Meinert et al., 1997) that contain substantial gold (>1 million ounces and > 1 g/t Au), but which are mined primarily for other commodities such as copper. It also excludes deposits such as the Veselyi Mine in the Siniukhinskoe District, CIS where gold was high-graded from a Cu-Au skarn system due to socioeconomic considerations, but which would have been mined for Cu-Au in most other societies (Ettlinger & Meinert 1992). Conversely, this definition includes deposits such as Bermejal (de la Garza *et al.* 1996) and Key East (Lowe & Larson 1996; Lowe 1998) that contain large amounts of other metals (such as Fe in the form of magnetite) that are not mined.

Reduced Gold Skarns

The highest grade (5-15 g/t Au) gold skarn deposits are relatively reduced, are mined solely for their gold content, lack economic concentrations of other metals, and have a distinctive Au-Bi-Te-As geochemical association. Most high-grade gold skarns are associated with reduced (ilmenite-bearing, Fe_2O_3/(Fe_2O_3+FeO) << 0.75) diorite-granodiorite plutons and dike/sill complexes. They typically occur in clastic-rich protoliths rather than pure limestone and skarn alteration of dikes, sills, and volcaniclastic units is common. Reduced gold skarns are dominated by iron-rich pyroxene (typically > Hd50), but proximal zones can contain abundant intermediate grandite garnet. Other common minerals include K-feldspar, scapolite, vesuvianite, apatite, and amphibole. Distal/early zones contain biotite±K-feldspar hornfels, that can extend for 100's of metres beyond massive skarn. Due to the clastic-rich, carbonaceous nature of the sedimentary rocks in these deposits, most skarn is relatively fine-grained.

Hedley District, British Columbia

The Nickel Plate mine in the Hedley district, British

Columbia is the largest and highest grade gold skarn in Canada. Discontinuous production from 1904 until the mine closed in 1995 was 13.4 million tons averaging 5.3 g/t Au, 1.3 g/t Ag, and 0.02 per cent Cu (Ray *et al.* 1996). Of this, more than 3 million tons of ore was mined underground at an even higher grade, averaging 14 g/t Au. Skarn formed in dominantly clastic rocks of the upper Triassic Nicola Group, that is part of the allochthonous Quesnel Terrane of the Intermontane Belt. Skarn is spatially and genetically associated with the dioritic Hedley intrusions, that comprise the Toronto Stock and a series of dikes and sills, many of which exhibit strong endoskarn alteration with abundant pyroxene, biotite, garnet, amphibole, and K-feldspar. Dating of these intrusions suggests an age range of 194-219 Ma (Ray & Dawson 1994). The Toronto Stock is a very reduced ilmenite-bearing intrusion with an average $Fe_2O_3/(Fe_2O_3+FeO)$ value of 0.15, the lowest of any gold skarn (Ray *et al.* 1995) and the lowest of any major skarn class (Meinert 1995).

As first recognised by Billingsley & Hume (1941), skarn is zoned in both space and time relative to the Toronto Stock and associated dikes and sills. The earliest and most distal alteration is a fine-grained biotite hornfels that affects both clastic rocks and some of the early sills (Ray *et al.* 1988). With time and proximity to massive skarn, biotite occurs with K-feldspar and pyroxene and is slightly coarser grained (Ettlinger 1990). This forms an aureole around the massive garnet-pyroxene skarn that is zoned from garnet pyroxene near the Toronto Stock to pyroxene-dominant (garnet:pyrox ene <0.1) skarn in distal ore zones (Ettlinger *et al.* 1992). Garnet is intermediate grandite in composition whereas pyroxene is relatively iron-rich. The most iron-rich garnet (Ad_{73-82}) occurs in distal ore zones (Ettlinger *et al.* 1992), whereas pyroxene composition changes systematically away from the Toronto Stock and the larger dikes, becoming more iron-rich and slightly more manganese-rich. The sulphide minerals associated with garnet and pyroxene skarn are dominantly arsenopyrite, loellingite, and pyrrhotite. Other sulphides, in decreasing order of abundance, are chalcopyrite, pyrite, sphalerite, hedleyite ($Bi_{2+X}Te_{1-X}$), native bismuth, gold, galena, and maldonite (Au_2Bi) (Ettlinger 1990). This latter group of minerals is mostly associated with lower temperature alteration including amphibole, ferroan wollastonite, scapolite, and prehnite. The scapolite and some amphiboles are unusually chlorine-rich and this feature has been suggested as an exploration guide to gold-rich systems (Pan 1998; Pan *et al.* 1994).

Garnet-pyroxene skarn at the Nickel Plate mine is estimated to have formed at a depth of 5 km and at an average temperature of 460-480°C, although fluid inclusions in some garnet and pyroxene samples homogenised at temperatures above 600°C (Ettlinger 1990). The presence of loellingite in addition to arsenopyrite also is suggestive of higher temperatures at a given sulphur fugacity (Kretschmar & Scott 1976, Heinrich & Eadington 1986). Salinities of garnet and pyroxene fluid inclusions average 18.3 and 9.7 wt. per cent NaCl equivalent, respectively, with sparse halite daughter minerals in garnet, pyroxene, and quartz yielding a maximum of 37.9 wt. per cent NaCl equivalent. Such high-salinity, high-temperature fluids would be capable of transporting significant gold as chloride complexes (Gammons & Williams-Jones 1995, 1997).

Fortitude Deposit, Battle Mountain District, Nevada

The Fortitude deposit is located in the Battle Mountain District of north-central Nevada and produced 77 Mt of Au from 10.9 Mt of ore at an average grade of 7.1 g/t Au (Doebrich & Theodore 1996). The Battle Mountain District contains several different skarn types ranging from a typical copper skarn with low gold grades, called the West Ore Body, to gold-rich, copper-poor skarns such as the Upper and Lower Fortitude ore bodies. The West ore body is adjacent to the 38-38.5 Ma (Theodore *et al.* 1978, K-Ar on primary biotite) Copper Canyon granodiorite porphyry. The skarns nearest the intrusive body are dominated by garnet with minor pyroxene and are typically high in copper with low concentrations of gold. The Fortitude deposit, like most high-grade Au skarns, has an unusual reduced skarn mineralogy and trace-element signature (Au-Bi-Te-As) that distinguishes it from most other skarn types. The reduced skarn mineralogy reflects the reduced nature of the associated Copper Canyon granodiorite [$Fe_2O_3/(Fe_2O_3+FeO) < 0.5$] that is quite distinct from typical oxidised porphyry copper deposit plutons. The more distal skarns contain more pyroxene than garnet and contain the highest concentrations of gold in the district (Myers 1994). Stylolites and other fluid escape structures are present in carbonate beyond the limit of calc-silicate formation.

Like much of central Nevada, detailed geologic relations in the Battle Mountain District are complicated by numerous thrust faults (De Witt, Golconda, and Roberts Mountain) that have juxtaposed a wide variety of rock types, all of which have been affected by hydrothermal alteration (Blake *et al.* 1984). Host rocks range from lower Cambrian through Permian and consist of sandstone, arkose, shale, chert, argillite, greenstone, limestone, and quartzite (Blake *et al.* 1984). Most skarn occurs in the basal mid-Pennsylvanian Battle Formation, the Pennsylvanian to Permian Antler Peak Limestone, and the Permian Edna Mountain Formation. The main mineralised unit is the Antler Peak Limestone that consists of well bedded dark gray limestone and minor chert nodules (Theodore & Blake 1978).

Skarn mineralogy has been investigated for most of the deposits in the Battle Mountain district (Theodore & Hammarstrom 1991; Doebrich *et al.* 1996). For example, the West Ore Body occurs mostly in the Antler Peak Limestone and is a typical copper skarn with a prograde skarn assemblage of grandite garnet (Ad39-99) + diopsidic pyroxene (Hd20-50) and a retrograde assemblage of actinolite + epidote + K-feldspar (Theodore & Blake 1978). Pyrrhotite, pyrite, chalcopyrite, and marcasite are the main sulphide minerals. The Fortitude deposit also occurs in the Antler Peak Sequence, primarily in the Antler

Peak Limestone and the Battle Formation, and can be subdivided into an upper and lower ore body due to offset across the Virgin Fault. The Fortitude ore body contains a prograde skarn assemblage of hedenbergitic pyroxene (Hd20-95, Jo<10) + andraditic garnet, with only minor retrograde alteration to epidote + actinolite + chlorite. Pyroxene shows a general increase in iron content moving towards the marble front (Hd30 proximal to the Copper Canyon stock and Hd>90 at the distal skarn fringe). This trend is mimicked by a Mn enrichment moving towards the marble front, with the pyroxene nearest the intrusion rarely exceeding 3 per cent Jo (except for crosscutting veinlets and crystal rims), whereas the pyroxene near the marble front is generally between 2-8 per cent Jo (Myers 1994).

The distribution of most metals parallels the skarn zonation in the Antler Peak Limestone. Copper is highest in garnet-rich skarn near the intrusive contact, whereas gold is concentrated in pyroxene-dominant skarn, particularly where the pyroxene is iron-rich ($>Hd_{50}$). Silver has elevated concentrations proximal to the stock and at the distal skarn front, beyond the main Fortitude gold zone (Kotlyar *et al.* 1998). The skarn system contains several sulphide species including chalcopyrite, pyrite, pyrrhotite, arsenopyrite, marcasite, sphalerite, and galena, that occur roughly in the order listed from intrusion to marble. Arsenopyrite is locally massive and native Bi is commonly visible in hand specimens. Native gold occurs at grain boundaries between skarn and sulphide minerals indicating a possible reaction relationship (Wotruba *et al.* 1988). In general, gold is associated with native bismuth, hedleyite, pearcite, and stannite. Trace elements are also zoned within the system with anomalous Co, Mo, Cr, and Ni in proximal zones and anomalous As, Bi, Cd, Mn, Pb, Zn, Sb, and Hg in distal zones.

Fluid inclusion work shows that the skarn formed at relatively high temperatures (300->550°C) that parallel fluid inclusion homogenisation temperatures measured in the adjacent Virgin dike apophysis of the Copper Canyon granodiorite (Myers 1994). The distribution of measured fluid inclusion temperatures parallels the skarn zonation. Garnet closest to the main stock (drill hole #500) ranges from 360-590°C. More distal garnet and pyroxene (drill holes 2723 and 1997) range from 380-440°C and 320-430°C, respectively and the most distal (and iron-rich) pyroxene (drill hole 1994) ranges from 350-400°C (Myers 1994). In addition, high salinity conditions have been documented, with multiple daughter minerals in fluid inclusions identified by SEM and STEM analysis. Limited fluid inclusion measurements indicate pyroxene skarn had salinities of 25-44 wt. per cent NaCl equivalent. Based upon limited evidence for boiling, Myers (1994) estimated a formation pressure of 0.4 kb (40 MPa) for the Fortitude system, in close agreement with the stratigraphic estimate of 1.5 km and a pressure of 375 bars (37.5 MPa) by Theodore & Blake (1975). As at Hedley, the presence of high-salinity, high-temperature fluids at Fortitude suggests gold transport by chloride complexes.

Similar zonation occurs in $\delta^{18}O$ and $\delta^{13}C$ values that indicate progressive reaction of a magmatic fluid with isotopically heavy carbonate wallrocks, as summarised by Zimmerman *et al.* (1992) and Myers (1994). Skarn garnets are progressively enriched in $\delta^{18}O$ outward from the Copper Canyon stock with garnet $\delta^{18}O$ values of 6.9 per mil in the proximal skarn and values as high as 8.2 per mil in distal skarn. Pyroxene ($\delta^{18}O$ = 8.6 to 10.3 per mil), amphibole ($\delta^{18}O$ = 8.6 to 9.2 per mil), and quartz ($\delta^{18}O$ = 11.4 to 13.2 per mil) are less systematic, but in each case the highest $\delta^{18}O$ values are most distal to the granodiorite stock. Skarn formation can be modelled as resulting from the progressive reaction of magmatic fluids with isotopically heavier carbonate wallrocks ($\delta^{18}O$ = 24.0 per mil). The variation in $\delta^{13}C$ values in calcite can also be explained by progressive reaction of magmatic fluids with carbonate wallrocks. $\delta^{18}O$ and $\delta^{13}C$ values decrease from unaltered limestone ($\delta^{18}O$ = 24.0 per mil, $\delta^{13}C$ = 2.4 per mil) to blocks of residual limestone in skarn ($\delta^{18}O$ = 15.4 to 19.3, $\delta^{13}C$ = -4.5 to 1.7 per mil) to calcite intergrown with skarn minerals ($\delta^{18}O$ = 11.8 to 13.1 per mil, $\delta^{13}C$ = -10.3 to -1.7 per mil). The absence of mineral phases with $\delta^{18}O$ less than magmatic values suggests that meteoric fluids ($\delta^{18}O$ <-10‰) probably did not play a significant role in the formation of this deposit (Zimmerman *et al.* 1992). This latter feature is consistent with the relatively small amount of retrograde alteration observed in the Fortitude skarn (Myers 1994).

Crown Jewel, Washington

Gold skarn mineralisation at the Crown Jewel deposit was discovered in 1988 and current reserves stand at 8.7 Mt averaging 6.0 g/t Au, for a total of 52 t Au, although the deposit is still in a protracted permitting phase (Bob Derkey, Washington Department of Natural Resources, personal communication). Skarn is most closely associated with the Cretaceous. Buckhorn Mountain granodiorite and a series of granodiorite porphyry dikes that have been interpreted by Hickey (1990) as cognatic. The main granodiorite pluton is quite reduced and contains primary ilmenite. It has a dioritic border phase that is more mafic and less silicic, but with similar alkalis relative to the central core. Hickey (1992) attributed the relatively high alkali content of the diorite border phase to alteration. The diorite is cut by garnet veins with pyroxene envelopes, although pervasive endoskarn alteration only occurs in some of the smaller dikes and sills.

The stratigraphy and structure of the host rocks at Crown Jewel are not well understood due to poor exposures and a regional metamorphic/shearing event that predates skarn formation (McMillen 1979). The shearing may be related to development of gneiss domes in the Okanogan highlands (Orr & Cheney 1987), although Hickey (1992) states that none of the skarn has been affected by shearing, *e.g.* there is no strain or deformation of skarn minerals. Rocks in the district that have been affected by alteration and mineralisation can be divided into distinct groups including a lower unit containing calcareous siltstone, sandstone, and minor shale; a limestone that has been converted to marble; an upper unit containing shale, minor siltstone, and sandstone; and a distinctive chert pebble conglomerate (Hickey 1990). These units are thought to correlate with the Paleozoic Anarchist Formation. Structurally overlying the Anarchist is the Permo-Triassic Kobau Formation consisting of andesitic volcanic rocks with shale and volcaniclastic interbeds.

Distal alteration, especially in argillaceous and clastic units, consists of biotite and pyroxene hornfels. Closer to intrusive contacts or fluid pathways these minerals become coarser grained and pyroxene replaces the biotite. In more calcareous rocks and limestone, the early/distal biotite and pyroxene hornfels are replaced by garnet. Some of the rocks behaved in a brittle fashion following pyroxene formation such that veins and breccias are cemented by brown garnet. Close to intrusive contacts, limestone is completely replaced by massive garnet and magnetite. This zonation is mirrored by an iron enrichment in pyroxene, with the most distal pyroxene approaching pure hedenbergite in composition. Retrograde alteration at Crown Jewel is relatively coarse grained and consists of epidote, amphibole, zoisite, calcite, and quartz. Sulphides are associated with retrograde alteration and with massive magnetite. Magnetite-pyrrhotite occurs as veins cutting garnet close to the granodiorite, as well as massive replacement of marble. In places the magnetite is abundant enough to have been mined on a very small scale in the past, although it is not of economic importance at present.

Pyrrhotite is the most abundant sulphide mineral by far, reflecting the overall reduced nature of the protolith, pluton, and skarn mineralogy. Other sulphides include pyrite, marcasite, chalcopyrite, bismuthinite, cobaltite, native gold, native bismuth, and arsenopyrite (Hickey 1990). Arsenopyrite is only abundant in the relatively impermeable and brittle chert pebble conglomerate. As with most reduced gold skarns, bismuth minerals are strongly associated with gold mineralisation. Crown Jewel may be unusual in that coarse grained bismuthinite is easily visible in drill core and is an excellent indicator of ore-grade gold (which is not visible at the hand specimen level). This bismuth-gold association is substantiated by assays of drill core composites.

Fluids associated with skarn formation and mineralisation at Crown Jewel were high-temperature saline brines. Hickey (1990) reports abundant, large halite daughter minerals in plutonic quartz, but did not find daughter minerals in the very small inclusions present in skarn minerals. Primary fluid inclusions in pyroxene homogenised from 365-450°C, whereas those in garnet, homogenised from 300-370°C. Two salinity determinations from fluid inclusions in garnet yielded values of 19 and 22 eq. wt. per cent NaCl. Fluid inclusions in epidote and amphibole yielded slightly lower homogenisation temperatures of 255-320°C and 315-350°C, respectively, for retrograde alteration.

Based upon an assumed depth of 4 km at the time of intrusion and skarn formation, Hickey (1990) determined an average lithostatic pressure corrected temperature for garnet-pyroxene skarn of 465°C. Quartz veins that cut garnet-pyroxene skarn have similar homogenisation temperatures with a wider range of salinity from 2-24 eq. wt. per cent NaCl.

Elkhorn, Montana

The Elkhorn district in Montana contains a variety of reduced gold skarns related to mafic diorite stocks marginal to the Boulder Batholith. Individual skarn deposits include Carmody, Diamond Hill, Dolcoath, East Butte, Elkhorn, Heagan, Glory Hole, and Sourdough. Historic production from skarns in the Elkhorn district was 2.1 tons of Au as a by-product of base-metal mining (Klepper *et al.* 1971). Recent exploration by several companies in the district has defined a combined resource of about 9 Mt averaging 4.8 g/t Au, based upon drilling of numerous discrete skarn zones (Everson & Read 1992 and unpublished abstracts). This represents a combined resource of more than 45 tons of gold contained in skarn.

The main phase of the Boulder Batholith is quartz monzonite. Satellite stocks at East Butte, Black Butte, and Cemetery Ridge stocks are dark, fine to medium-grained diorites which are similar in age to slightly older than the Boulder Batholith (Everson & Read 1992). These plutons have intruded a lower Paleozoic sequence including the Wolsey, Meagher, Park, Pilgrim, Maywood, Red Lion, Jefferson, Three Forks, and Madison formations. Near plutons, argillaceous rocks of the Park, Wolsey, and Three Forks formations have been converted to biotite, pyroxene, and calc-silicate hornfels, similar to that described at many other gold skarns, whereas the generally dolomitic carbonate units of the Meagher, Pilgrim, Maywood, Red Lion, Jefferson, and Madison formations have been recrystallised and locally silicified.

Skarn associated with the East Butte Diorite occurs as endoskarn in the diorite and as exoskarn in two stratabound units near the Wolsey-Meagher contact, which strikes NNW and dips 60-70°E (Everson & Read 1992). Exoskarn consists of dark green pyroxene and minor garnet. Pyrite, pyrrhotite, magnetite, and arsenopyrite occur disseminated in skarn, averaging 3-5 per cent, and as massive replacement zones near the marble front. Minor phases recognised petrographically include marcasite, maldonite (Au_2Bi), hedleyite ($Bi_{14}Te_6$), hessite (Ag_2Te), gersdorfitte (NiAsS), and native bismuth (Meinert unpublished data). Retrograde alteration consists of amphibole, phlogopite, vesuvianite, and epidote. About half of the mineralisation occurs as endoskarn alteration of the East Butte Diorite (Everson & Read 1992). Endoskarn consists of pyroxene, calcic plagioclase (close to pure anorthite), amphibole, titanite, and local veins of quartz-orthoclase.

Near the historic Carmody mine (Klepper *et al.* 1971), skarn associated with the East Butte Diorite occurs as a single stratabound layer in the Wolsey Formation. Skarn is presumed to have replaced a carbonate layer and is surrounded by biotite hornfels in the originally more argillaceous lithologies. Carmody mineralisation is similar to the East Butte mineralisation except pyrrhotite is more abundant than pyrite, and both chalcopyrite and sphalerite occur in minor amounts.

Significantly different skarn mineralisation occurs in the Sourdough zone northwest of East Butte near the historic Golden Curry mine (Roby *et al.* 1960). Sourdough skarn is spatially associated with both monzonite and diorite and occurs as endoskarn within the monzonite and as replacement of dolomitic rocks thought to be either Pilgrim or Jefferson Formation (Everson & Read 1992). Both endoskarn and exoskarn are pyroxene dominant with little garnet. Massive magnetite occurs at the marble front and within exoskarn associated with pyroxene, olivine, ludwigite-vonsenite ($(Mg,Fe)BO_5$), and phlogopite. Retrograde alteration consists of abundant serpentine and tremolite.

Junction Reefs, Australia

Although large scale production is relatively recent, the gold skarns at Junction Reefs, New South Wales, Australia have been mined since 1876 with historical production of 1.1 tons of Au between 1876 and 1938 (Gray *et al.* 1995). Open pit mining began in 1988 and continues to the present, with total skarn reserves and production of 2.4 Mt with an average grade of 3.3 g/t Au, representing 7.7 tons of Au.

The protolith for skarn mineralisation at Junction Reefs is a 39 m thick sequence of marine limestone, siltstone, and chert that occurs within the voluminous (>2500 m). Early Ordovician Coombing Formation consisting of massive volcanic graywacke, cherty shale, siltstone, and tuffaceous arenite (Gray *et al.* 1995). Like most turbidite sequences in an island arc setting, there are intercalations of volcanic flows and tuffs, but some workers regard the overall tectonic setting as one of shallow basins overlying a thin continental crust (Wyborn 1988). Intrusive into the Coombing Formation are a series of shoshonitic diorites, monzodiorites, monzonites, and quartz monzonites. In the

Junction Reefs district numerous, locally interconnected, monzodiorite stocks, dikes, and sills were intruded between 430 and 440 Ma (Gray *et al.* 1995).

The Junction Reefs monzodiorite is surrounded by a zoned skarn system which has ore grade gold mineralisation in the outer zones. Because most of the Coombing Formation consists of relatively unreactive siliciclastic rocks, skarn formation and mineralisation are restricted to stratigraphic/structural windows of more calcareous rocks within the metamorphic/hydrothermal aureole of the monzodiorite and associated dikes and sills. However, as in many other gold skarn districts, the siliciclastic rocks have been converted to purple-brown biotite hornfels within 200 metres of the Junction Reefs monzodiorite (Gray *et al.* 1995). Closer to the intrusion and along bedding planes, fluids forming amphibole and pyroxene have infiltrated the rock, forming a green biotite-amphibole-pyroxene hornfels. This rock is not visually striking except when split open along bedding planes to expose radiating clusters of dark green amphibole and pyroxene crystals with interstitial diamond-shaped arsenopyrite up to 1 cm in length.

The skarn system is zoned around the Junction Reefs monzodiorite and three separate mines (Sheahan-Grants, Frenchmans, and Cornishmens) occur where calcareous rocks are exposed in the outer skarn zones. The inner most skarn zone (termed Zone 1) consists dominantly of pale green garnet, lesser quartz, and 20 per cent pyroxene. Both garnet and pyroxene range up to the pure Fe end members. Minor pyrite (5 per cent) is disseminated in the garnet skarn and gold grades are low, averaging 0.1-0.2 g/t Au (Gray *et al.* 1995). In Zone 2 pyroxene is much more abundant than garnet and is slightly more iron-rich, on average, than in Zone 1, ranging from ferrosalite to hedenbergite (Hope 1990). Minor chlorite is reported with pyrrhotite and pyrite (Grant 1988). In Zone 3 the prograde pyroxene garnet skarn has been strongly overprinted by amphibole approaching ferrohastingsite in composition. Pyrrhotite is the dominant sulphide and is associated with amphibole, and to a lesser extent, with chlorite, calcite, and quartz. Gold reaches ore grade (1.0 g/t Au) locally within Zone 3. Zone 4 hosts the vast bulk of ore grade mineralisation. Remnant textures of garnet and pyroxene are present and rare small grains survive armoured in quartz or sulphide, but most Zone 4 rocks are a dark green felted mass of chlorite, calcite, quartz, and sulphides. The dominant sulphide is pyrrhotite with lesser arsenopyrite, chalcopyrite, pyrite, and marcasite. Minor phases include native bismuth, maldonite, and an unidentified Au-Bi sulphide mineral. Zone 4 averages 10-20 per cent sulphide and 80 per cent massive sulphide occurs locally at the marble front. Gold is associated with sulphides and high concentrations of arsenopyrite (core assays range from 0.01-9.55 per cent As) correlate with very high gold grades (Gray *et al.* 1995). Locally, there is a zone of wollastonite, vesuvianite, quartz ± grossularitic garnet at the marble front. This has been designated Zone 5, but is not as continuous as the other four zones.

Geochemically, skarn at Junction Reefs is anomalous in Au, As, Bi, Co, Fe, Pb, and Zn. As with many other gold skarns, the strongest correlation ($r = 0.83$) is between Au and Bi. Au and As are only moderately correlated ($r = 0.58$) and most other elements do not exhibit a systematic correlation with Au (Gray *et al.* 1995). Even in ore zones, Ag is very low, < 3 ppm. Fluid inclusions have not been examined in calc-silicate minerals from Junction Reefs. However, fluid inclusions in quartz and calcite have homogenisation temperatures up to 345°C and 325°C, respectively and salt daughter minerals were observed in some inclusions, indicating at least some fluid salinities > 26 eq. wt. per cent NaCl (Grant 1988). These temperatures are in broad agreement with those determined for retrograde alteration at Junction Reefs from chlorite geothermometry (Grant 1988), thus indicating a minimum temperature for the system.

Beal, Montana

The Beal deposit is located approximately 26 km west-southwest of Butte, Montana and has proven and probable ore reserves of 14.8 Mt at an ore grade of 1.49 g/t Au, totalling 23.1 tons of Au. The deposit is hosted by late Cretaceous clastic, fluvial-deltaic sedimentary rocks of the Vaughn member of the Blackleaf Formation (Wilkie 1996). In the vicinity of the Beal mine, the Blackleaf Formation has been metamorphosed and metasomatised to a peak grade of pyroxene hornfels by diorite and granodiorite intrusions (74.8 + 2.8 m.y., K-Ar date on biotite) related to the Boulder Batholith (Hastings & Harrold 1988). A K-Ar date (71.7 ± 2.6 m.y.) on adularia in a gold-bearing vein at Beal suggests that mineralisation and intrusion are closely related (Hastings & Harrold 1988).

Structurally, the Beal deposit lies approximately three kilometres east of the western margin of the frontal fold and thrust belt of southwestern Montana. This zone is marked by a series of north-south trending thrust faults (Johnson, Spring, and Long Tom Thrusts), which juxtapose older (Paleozoic and Pre-cambrian) rocks over the Cretaceous Blackleaf Formation (Ruppel *et al.* 1981). Thrusting predates the Beal deposit and is unrelated to mineralisation. Numerous steep faults cut the Beal deposit, the most important being the German Gulch fault, Beal shear zone, and Gully fault. The Beal shear zone trends N80-85°W and dips 85-90°S, is locally mineralised, and was an important structural control for channelling hydrothermal fluids (Wilkie 1996).

All known mineralisation at the Beal deposit occurs within the hornfels aureole of the granodiorite and diorite intrusions. Granodiorite of the Boulder Batholith crops out along the eastern edge of the mine area and numerous small dioritic stocks and dikes crop out near the margin of the batholith and within the open pit mine. Diorite samples from the pit are dark greenish-gray to greenish-black and consist of fine- to medium-grained plagioclase, biotite, amphibole, pyroxene, and K-spar. Opaque mineralogy consists of pyrite, chalcopyrite and magnetite-ilmenite intergrowths. The presence of ilmenite is indicative of a reduced magma chemistry and may be related to the gold content of this igneous system. All of the diorite exposures in the pit are intensely altered and contain up to 15-20 per cent hydrothermal biotite/chlorite alteration of primary pyroxene, hornblende, biotite, and feldspar (Wilkie 1996).

Samples from traverses extending 3 km E-W perpendicular to the intrusive contact demonstrate mineralogical and temperature zonation outward from the pluton as follows: granodiorite → pyroxene → amphibole → biotite → white mica (Wilkie 1996). The width of the mineralogical zones is approximately constant throughout the area and within each zone mineral abundance decreases regularly (for a given protolith) with distance from the pluton. A marked exception is the high abundance of pyroxene near the Beal mine. Scapolite (containing 2-3 wt. per cent Cl) is also abundant in this area. Biotite, chlorite, and sericite geothermometers [models of McDowell and Elders (1980) and Walshe (1986)] indicate a temperature decrease away from the intrusive contact with a thermal anomaly coincident with the pyroxene-scapolite zone in the traverse which passes through the ore deposit (Wilkie 1996).

Pyroxene in pale green pyroxene hornfels occurs as <0.1 mm equant grains or clusters in the matrix among quartz grains. Most pyroxene compositions are from Hd_{18-40}. Close to the Beal mine, pyroxene is slightly coarser grained, darker green, and more Fe-rich (Hd_{42-62}). Coarser-grained pyroxene occurs along fractures and at fracture intersections as orbicular patches. Even in these occurrences pyroxene never constitutes a majority of the rock. Thus, alteration at Beal has far less pyroxene and overall is far less intense (garnet is absent) than any of the other reduced skarns considered in this review. This is considered to be a direct result of the carbonate-poor protolith and the relatively low temperature of skarn formation (Wilkie 1996).

Amphibole at Beal is compositionally more complex than pyroxene and ranges from calcic tremolite-actinolite to hornblende and ferrohastingsite. The iron-rich nature of the amphibole is surprising in light of the iron-poor nature of the host rocks. Locally, amphibole occurs as a replacement of pyroxene, but in most cases there is not textural or compositional evidence for pre-existing pyroxene (Wilkie 1996). Thus, amphibole at Beal appears to be a primary alteration phase. Strongly mineralised rocks can contain more than 50 per cent amphibole and the amphibole typically is associated with K-feldspar, chlorite, and sulphides.

The main sulphide phases include pyrrhotite, pyrite, chalcopyrite, arsenopyrite, and minor tetrahedrite. Overall, Beal is a sulphide-poor system. Sulphide contents typically are less than 2 per cent and many areas of ore grade material contain less than 1 per cent total sulphide (Wilkie 1996). Local areas of $\geq$ 5 per cent arsenopyrite and/or pyrrhotite occur but are not always associated with high gold grades. Gold mineralisation occurs as 1-5 micron grains disseminated in coarser-grained meta-sedimentary rocks and as flakes in quartz and quartz-sulphide-adularia veins (Hastings & Harrold 1988). The quartz-sulphide-adularia veins occur in steep structures within the pit and cut both the hornfels and plutons. Geochemical analyses of Cu, Bi, Zn, As, Au, and Ag in run-of-mine ore show that Au strongly correlates with Bi, weakly correlates with As and Cu, but does not correlate with other elements (Wilkie 1996).

Due to the very small grain size of most calc-silicate minerals at Beal, fluid inclusion studies have been limited. However, Wilkie (1996) measured fluid inclusions in quartz from several alteration styles at Beal. Measured homogenisation temperatures range from 115-550°C with averages for L-V inclusions, L-V-salt, and vapour-rich inclusions of 344°, 312°, and 359°C, respectively. Salinities ranged from 4.7-42 eq. wt. per cent NaCl; the salinity of fluid inclusions with daughter minerals ranged from 34-42 eq. wt. per cent NaCl. Evidence for boiling in association with Au-Bi-Te minerals was documented at 340°C and salinities of 5-20 eq. wt. per cent NaCl. Similar temperatures were determined from mineral composition geothermometers, including arsenopyrite

(408-428°C), biotite (362-400°C), and chlorite (314-378°C). In addition, both biotite and chlorite geothermometers indicate an overall temperature decrease away from the main intrusive contact.

$\delta^{18}O$ values for both coarser-grained (13-15.9 per cent) and finer-grained (12.9-15.5 per cent) siltstone samples at Beal increase away from the Boulder Batholith; coarser-grained samples have $\delta^{18}O$ values that are generally 0.5 higher than finer-grained samples for a given location (Wilkie (1996)). This distribution is a function of two interacting processes: 1) decreasing temperature and 2) decreasing water-rock ratio away from the contact. $\delta^{34}S$ isotopic values are tightly clustered for a given mineral: pyrrhotite (7.7-9.3 per cent), pyrite (7.0-8.2 per cent), chalcopyrite (7.6-9.1 per cent), arsenopyrite (8.8-14.0 per cent), galena (9.2), and sphalerite (13.0 per cent). In most cases, fine-grained disseminated sulphides have lower $\delta^{34}S$ isotopic values than vein sulphides and coarse-grained crystals. There is no evidence for a sedimentary sulphur reservoir; most samples are consistent with a dominantly magmatic source of sulphur (Wilkie & Meinert 1994).

Oxidised Gold Skarns

Whereas the "classic" gold skarn deposit is characterised by low garnet:pyroxene ratios, hedenbergitic pyroxene, and abundant sulphides dominated by pyrrhotite and arsenopyrite, several skarns have been mined for gold that have a very different mineralogy and mineralisation style. These deposits have been classified by Brooks *et al.* (1991) as oxidised gold skarns.

Their essential features include high garnet:pyroxene ratios, relatively Fe-poor garnet and pyroxene, low total sulphides, pyrite≥pyyrhotite, and minor but ubiquitous occurrences of chalcopyrite, sphalerite, and galena. In addition, the highest gold grades are not associated with prograde garnet-pyroxene, but rather with later retrograde alteration including abundant K-feldspar (adularia) and quartz. Some of these deposits can be considered transitional to other types of gold mineralisation such as epithermal deposits in which phase separation (boiling) can be an important precipitation mechanism (*e.g.* Hedenquist *et al.* 1996).

McCoy, Nevada

The McCoy gold skarn is only 45 km southwest of the reduced Fortitude gold skarn in northcentral Nevada, but differs dramatically in regards to the style of mineralisation and wallrock alteration. The McCoy deposit contains 15.6 Mt of ore averaging 1.44 g/t Au and an additional 30,430 tonnes averaging 14.6 g/t Au that was mined underground (Brooks 1994). Production is from garnet-rich skarn surrounding the 39 Ma Brown stock, a reduced ilmenite-series, hypabyssal, hornblende-biotite granodiorite. Brooks (1994) subdivided the Brown stock into five petrologically distinct phases and invoked mixing of discrete magmas to yield individual intrusive phases. Importantly, there are systematic correlations between individual intrusive phases and the mineralogy and gold grade of associated skarn. The Brown stock is estimated to have intruded to within 1.3 km of the surface and this shallow emplacement is reflected by the multitude of dikes and sills found on the margins of the main stock. In addition, most of the early dikes and sills have been affected by garnet-pyroxene endoskarn.

Skarn at McCoy is zoned in both space and time. The earliest and most distal alteration is biotite and pyroxene hornfels. This results in a pale, fine-grained rock with original sedimentary layering still preserved. Overprinting this hornfels are veins and massive zones of garnet-dominant skarn. Typical garnet:pyroxene ratios are 3:1 to 20:1. Close to intrusive contacts, all the hornfels has been replaced and no trace of sedimentary bedding is left. Skarns closest to the main intrusion, called the West Contact and Peacock skarns, are the only skarns with significant pyroxene (>10 per cent), and also the only pyroxene that is relatively coarse-grained and Fe-rich (up to Hd_{75}). All other skarn at McCoy is garnet dominant and where pyroxene is present, it is diopsidic. Early garnet is Fe-poor and occurs as bedding replacements of argillaceous layers (skarnoid) and as cores to later metasomatic garnets, that are more Fe-rich. These compositional differences are important in that subsequent retrograde alteration selectively replaces certain stages and compositions of garnet and pyroxene (Brooks 1994). Sulphide minerals associated with prograde skarn include pyrrhotite, pyrite, sphalerite, galena, arsenopyrite, chalcopyrite, bornite, gold, hedleyite, native bismuth, and hessite (Brooks 1994).

Late garnet-pyroxene skarn coexists with or is overprinted by retrograde alteration consisting mainly of epidote-quartz-pyrite-K-feldspar. Grandite garnet is more susceptible to retrograde alteration than is andradite garnet. Biotite and chlorite occur instead of epidote in distal zones of retrograde alteration and where pyroxene was relatively abundant. Most economic gold mineralisation is associated with retrograde alteration, particularly with quartz-pyrite-K-feldspar. The K-feldspar varies in colour from pink to a pale tan and is similar to the adularia described from many epithermal deposits. The most intense quartz-pyrite-K-feldspar is spatially associated with a particular generation of dikes and sills called the Productive Series (Brooks 1994). However, quartz-pyrite-K-feldspar also replaces distal skarn and locally occurs as silicified pods in limestone beyond the limit of garnet-pyroxene alteration. This latter occurrence is similar to the jasperoids associated with some epithermal gold deposits, particularly Carlin-type deposits.

The fluids associated with prograde garnet and pyroxene at McCoy are high-temperature brines. Brooks (1994) reported homogenisation temperatures in garnet ranging from 330-590°C with an average of 493±46°C. Measured salinities ranged up to 39.8 eq. wt. per cent NaCl. Homogenisation temperatures in pyroxene range from 300-420°C and the average for proximal pyroxene is 398±14°C, whereas the average for distal pyroxene is 322±14°C. This spatial decrease in temperature is mirrored by a decrease in salinity. The salinity of fluid inclusions in proximal pyroxene ranges up to 35.3 eq. wt. per cent NaCl, whereas the maximum measured salinity in distal pyroxene is 22 eq. wt. per cent NaCl.

The fluids associated with retrograde alteration are slightly lower in temperature and salinity than those measured in prograde skarn, but are well above values typically reported for epithermal systems. Fluid inclusions in epidote (which replaces garnet) range from 360-450°C with salinities up to 28 eq. wt. per cent NaCl. Both the temperature and salinity of fluid inclusions in epidote are less than the values measured in garnet. Fluid inclusions were also measured in quartz and K-feldspar associated with retrograde alteration. Fluid inclusions in vein quartz range from 280-360°C with salinities from 11-19 eq. wt. per cent NaCl. Fluid inclusions in K-feldspar range from 160-390°C with salinities from 17-32 eq. wt. per cent NaCl.

Brooks (1994) estimated a pressure of 350 bars for skarn formation at McCoy and used this to determine an average pressure correction of 30°C for the measured homogenisation temperatures. Collectively, these data indicate that prograde skarn formed at 330-620°C from brines with salinities up to 35 eq. wt. per cent NaCl. As temperatures declined, the early-formed garnet and pyroxene were altered to lower temperature assemblages including epidote, quartz, and K-feldspar. These minerals also formed from saline brines, but at temperatures 100-200°C lower than the prograde garnet and pyroxene.

Nambija, Ecuador

Ecuador has two significant gold-bearing skarns, Ximena and Nambija. Ximena in west-central Ecuador is a typical reduced gold skarn similar to Hedley and Fortitude in North America. It has produced about 75,000 ounces of gold from alluvial fields developed from a pyroxene-dominant skarn. In contrast, Nambija in southeastern Ecuador is an oxidised gold skarn with similarities to the McCoy skarn in Nevada and Red Dome in Australia. Its mineralogy is dominated by grandite garnet and most production has come from alluvial workings and high-grading by local campesinos. Nambija may be best known for spectacular colour photographs in the popular press that illustrate an ant swarm of human workers in the open pits reminiscent of gold rush days of previous centuries. Geologically, less is known about Nambija than most other gold skarn deposits due to the lack of organised mapping and the "unsettled" property ownership situation relative to the surface workings.

Nambija is one of a series of gold deposits in the southern portion of the Cordillera Real, a north-northeasterly trending belt of Cenozoic, Mesozoic, and Paleozoic rocks. The central part of this belt consists of Tertiary to recent volcanic rocks, with several active volcanoes. West of this volcanic belt is an accreted Cretaceous sequence of island arc and oceanic sedimentary, volcaniclastic, and volcanic rocks, that have been intruded by numerous Tertiary I-type, relatively mafic plutons. This belt hosts the Ximena gold skarn deposit. East of the central volcanic belt lies a series of Paleozoic metamorphic rocks and Mesozoic sedimentary-volcanic rocks, which have been deformed by a fold and thrust belt. Along the general contact between the Paleozoic and Mesozoic belts are several large Jurassic plutons and the Nambija deposit is located in a pendent in one of these batholiths.

On a regional scale, the Nambija district is dissected by west-verging, N10°E to N20°E thrust faults spaced approximately 10-30 km apart. The Nambija skarn deposits occur within metamorphosed Piuntza volcano-sedimentary rocks that occur as roof pendants in the 170 Ma Zamora batholith (Litherland *et al.* 1994). The Piuntza Unit is approximately 500 m thick and consists of sandstone, siltstone, limestone, tuff, and andesitic flows (Paladines & Rosero 1996). The Zamora batholith is an equigranular tonalite to granodiorite (Salazar 1988). Other igneous rocks that have been reported in the Nambija district include monzodiorite, monzonite, rhyodacite, syenite, and quartz-feldspar porphyry dikes and small stocks (Hammarstrom 1992; Paladines & Rosero 1996). However, most of these intrusions have been altered to K-feldspar, sericite, chlorite, and clay. Thus, the original compositions and ages of these intrusions are not well known.

Within the Nambija district, there are a series of gold-bearing skarns, that have been worked by local campesinos, including from north to south, Fortuna, Campana, Campanilla, Nambija, Guaysimi, and Sultana del Condor. Artisanal workings at Nambija are estimated to have produced 2 million ounces of gold and the current resource is estimated at 23 Mt (Mining Magazine 1990). Reported grades range from 14 to 84 g/t Au, with an average of 15-30 g/t Au (McKelvey 1991; Hammarstrom 1992). Campanilla and Campana are smaller but of similar grade (Mining Magazine 1990). Given the coarse grain size of the gold and the rudimentary nature of the alluvial and artisanal workings, all of the above tonnage and grade figures should be viewed with caution. Most skarn pockets and mineralised zones occur in a north-northeast structural corridor of breccias, veins, and shears that parallel the larger faults.

This mineralised zone is 1.5 km long, 125 m wide, and dips 34°E within the pendent (Aguirre *et al.* 1985; McKelvey 1991). The highest-grade mineralisation occurs at the intersection of these northerly structures and northeast striking faults. Where these intersecting fault zones cut skarn, the rock is dissected by parallel quartz stringers with native gold and few if any sulphide minerals (Aguirre *et al.* 1985). The fact that most of the mineralisation and some of the skarn is structurally controlled and spatially associated with porphyritic rocks suggests that skarn formation and mineralisation are not related to the main phase of the Zamora granodiorite. Instead, skarn formation appears to be associated with some of the younger porphyritic intrusions and mineralisation is associated with quartz stringers that have a strong structural control.

There is a stock of quartz monzonite or rhyodacite porphyry at Nambija in the Tierrero 2 mine. The stock is surrounded by green garnet skarn with a zone of pink K-feldspar flooding and brecciation to the Southwest. The skarn is not sulphide-rich, but most samples contain minor pyrite, chalcopyrite, sphalerite and/or galena. In hand specimen, both the garnet and pyroxene are pale green in colour. In addition, some of the garnet has pale brown and yellow hues as well. Such pale green-yellow garnet is typical of distal skarn and is similar to the garnet in many Zn skarns. In thin section, the garnet is strongly zoned as is typical of hydrothermal skarn garnet. There are discrete cores and rims to most grains indicating multiple pulses of hydrothermal fluid and in general, rims are more andraditic than cores, *e.g.* normal zoning. Almost all the garnet analyses reported from these rocks range from Ad_{21-72} except for a few distal samples with pure andradite. Although not highly anomalous, most of the garnets contain 0.5-1.5 per cent MnO.

This is slightly more spessartine component that would typically occur in Au skarn garnets. Otherwise, these intermediate grandite compositions are typical of Au skarns and would be quite unusual for most base metal skarn systems, including Fe, Cu, and Zn-Pb (Meinert 1992). In contrast, all the pyroxene is diopsidic and such iron-poor pyroxenes are atypical of Au skarns. The pyroxenes also are relatively manganese-rich (Hd_{16-34} Jo_{5-13}), more than any other reported Au skarn, but significantly less than typical Zn skarns. The combination of high garnet:pyroxene ratios and both iron-poor garnet and pyroxene suggests that the Nambija system is both oxidised and iron-poor. This is consistent with the mineral abundances, compositions, and relative lack of iron sulphide minerals.

Gold at Nambija occurs with quartz veins spatially associated with garnet skarn. Some of the quartz veins have garnet envelopes indicating general contemporaneity with skarn formation. The fluid inclusions in the quartz are simple two phase inclusions. There are no daughter minerals, so the total salinity is < 26 wt. per cent NaCl. Homogenisation temperatures were not determined, but the lack of retrograde reaction with garnet, such as the formation of epidote, suggests that the temperature of quartz veining is relatively high and beyond the range of epithermal-type mineralisation.

There appears to be a transition from quartz veins with garnet envelopes to quartz veins and quartz flooding of the rock with no apparent reaction. Again, the lack of retrograde reaction with garnet, such as the formation of epidote, suggests that the temperature of quartz veining is relatively high. At the Campana mine, brown garnet skarn is cut by parallel quartz veins with a sheeted/ribbon texture. This rock clearly records two separate events.

The first event is the formation of relatively coarse-grained garnet skarn with optical zonation similar to other Nambija samples (core composition Ad_{40}, rim Ad_{60}). Pyroxene in this sample has a similar iron content to the other Nambija samples, but the manganese content is even higher than the other samples ($Hd_{31}Jo_{13}$). The second event is a brittle deformation in which the rock has been veined by hundreds of parallel quartz veins. The walls of the quartz veins match perfectly, requiring that brittle fracture occurred without significant shear. The garnet crystals have been sliced, as if by a comb, into dozens of parallel slivers, with each sliver separated by optically continuous quartz. In both the quartz veins and quartz flooding there is no apparent reaction of the hydrothermal fluids with the wallrock (garnet).

The fluid inclusions in the quartz are mostly vapour-rich indicating that boiling/fluid exsolution has occurred, probably due to a sudden pressure reduction (caused by fault movement). There are no daughter minerals, so the total salinity is < 26 wt. per cent NaCl. This texture is similar to that observed in mesothermal orogenic gold deposits where quartz veins contain tens to hundreds of ribbons of sheared wall rock, separated by quartz.

Magnesian Gold Skarns

Although one million tons at an average grade of 6 g/t Au

was produced from magnesian skarn at the Cable Mine, Montana (Earll 1972), most gold skarns are calcic skarns and little has been published until recently on the occurrence of magnesian gold skarns (Ettlinger *et al.* 1996; Mueller 1997). Most magnesian skarns form from dolomitic protoliths and exhibit a diagnostic mineralogy that includes forsterite, spinel, and serpentine. Although a variety of spinel phases can be present, magnetite usually is dominant and thus, most magnesian skarns are mined for iron and are relatively easy to find due to their strong magnetic signature. Butte Highlands, in southwest Montana is an unusual magnesian skarn in that it is an important gold resource, but lacks abundant iron oxides and sulphides. As pointed out by Ettlinger *et al.* (1996), the Fe-poor nature of this deposit means that it, and others like it, may not stand out during standard geophysical surveys.

Butte Highlands, Montana

Butte Highlands is one of many gold-bearing skarns associated with the relatively mafic marginal intrusions of the Boulder Batholith. Butte Highlands is located on the southern margin of the Boulder Batholith, about 24 km south of Butte, Montana. Skarn is associated with a fine- to medium-grained equigranular diorite, that has been intersected in drill core beneath the main mineralised area, called Nevin Hill (Ettlinger *et al.* 1996). Close to contacts with sedimentary rocks, the diorite exhibits endoskarn alteration with hornblende replaced by diopsidic pyroxene and titanite, plagioclase replaced by zoisite and prehnite, and calcium enrichment of plagioclase (to bytownite). In addition, the diorite is cut by veinlets of pyrrhotite with orthoclase, tremolite, and calcite envelopes.

The diorite has intruded the lower Paleozoic stratigraphic section at Butte Highlands causing extensive hornfelsing and recrystallisation of the Wolsey, Meagher, Park, and Pilgrim formations. Argillaceous rocks of the Park Formation have been converted to biotite and pyroxene hornfels, similar to that described at many other gold skarns, whereas the dolomitic Meagher and Pilgrim formations have been recrystallised and locally silicified (Ettlinger *et al.* 1996). Mantos and chimneys of massive sulphide replacement ore in the Meagher and Pilgrim marbles were mined for base metals in the early part of the century (Sahinen 1950), but the bulk of skarn and gold mineralisation occurs in the Wolsey Formation and the base of the Meagher Formation. The Wolsey Formation at Butte Highlands is described by Ettlinger *et al.* (1996) as consisting of interlayered, non-fossiliferous, dolomitic mudstone and shale, with some units of siltstone and carbonate.

Prograde skarn at Butte Highlands is dominated by forsteritic olivine with lesser pyroxene and phlogopite. Although these minerals are all pale green in colour, this rock is black in hand specimen due to pervasive serpentisation. Garnet is not abundant at Butte Highlands, but is present in endoskarn and with spinel as an overprint of the earlier olivine skarn. Such overprinting of early magnesian skarn minerals by later calcic skarn minerals has been reported from numerous magnesian skarn systems worldwide (Aksyuk & Zharikov 1988; Pertsev 1991).

Neither the garnet nor spinel are Fe-rich, in contrast to most skarn systems. Retrograde alteration of olivine results in abundant serpentine, phlogopite, talc, carbonate, and magnetite. Retrograde alteration of more calcic skarn results in amphibole and vesuvianite, minerals that contain both Mg and Ca. Sulphide mineralisation is strongly associated with retrograde alteration and Ettlinger *et al.* (1996) identified two associations with gold: phlogopite+pyrrhotite+gold and chlorite+clay+pyrrhotite+gold. In addition to this mineralogical association of gold with retrograde alteration, there is an elemental association of Au with Bi, based upon drill core assays (Ettlinger *et al.* 1996).

Skarn in "Mesothermal" Regional Metamorphic Terrains

Most skarns are associated with relatively shallow Phanerozoic plutons that have intruded previously unmetamorphosed sedimentary rocks (*e.g.*, Einaudi *et al.* 1981).

However, skarn mineralogy also has been described from several deposits in older orogenic belts where skarn is associated with both plutonism and high T-P metamorphism (*e.g.*, Lucky Draw, Australia, Sheppard *et al.* 1995; Navachab, Namibia, Noertemann 1997; Tillicum, British Columbia, Ray *et al.* 1986a; Peterson 1996). In addition to these plutonic/metamorphic occurrences, there are several "mesothermal" lode gold deposits with skarn alteration in Pre-cambrian terranes without associated intrusive rocks.

These occurrences are significantly different from Phanerozoic skarn systems and little is known about the geologic relations of the skarn alteration or the connection between gold mineralisation and skarn formation. Many researchers are unaware that these skarn occurrences even exist and there is much uncertainty about the timing and geochemistry of skarn formation. These skarns appear to be hybrids with characteristics of both the regional metamorphic environment and more typical Phanerozoic plutonism. What unites these disparate occurrences is a mineralogy dominated by very Fe-rich and reduced assemblages including garnet with major almandine-spessartine, hedenbergitic pyroxene, and Fe-rich amphibole. In some cases it appears that an Fe-rich protolith such as iron formation, komatiite, or metabasite is responsible for the unusual mineralogy. In addition, these deposits typically have part or all of the Au-As-Bi-Te geochemical signature of the younger gold skarn deposits. These "metamorphic" deposits are presented as a group because of their common link to regional metamorphism, even though there are huge differences in geologic setting and geochemistry among them. As more deposits like these are identified, it is hoped that understanding of their characteristics and origin will increase.

Lucky Draw, Australia

The Lucky Draw mine is located in the Burraga district 150 km west of Sydney in the Paleozoic Lachlan fold belt. Rocks in the Burraga district have been affected by two episodes of folding resulting in a series of upright, north-trending D_1 anticlines and synclines. D_2 folds are associated with Devonian-Carboniferous upper greenschist metamorphism and a regional slatey cleavage estimated to have formed at P_{total} = 2.0-2.5 kb (200-250 MPa) and T = 470° ± 35°C (Fowler 1987, 1989). Synchronous with regional metamorphism, a series of granitic plutons were emplaced with contact aureoles containing andalusite and cordierite. One of these, the Bathurst Granite, 30 km north of Lucky Draw, has been dated at 310 ± 7 Ma (Andrew 1984).

The Lucky Draw mine is located in the west-dipping limb of the D_1 Brownlea anticline close to the contact with the Burraga Granodiorite. Mineralisation occurs in the uppermost 100 m of the Ordovician Triangle Group, which consists of micaceous quartzite and quartz-mica schist, containing the assemblages quartz-albite-biotite-muscovite-cordierite and quartz-biotite-muscovite-albite-andalusite-cordierite, respectively (Sheppard *et al.* 1995). Within a few metres of the Burraga Granodiorite these assemblages are replaced by the assemblage quartz-biotite-plagioclase-cordierite-and alusite-sillimanite-Kfeldspar. Overlying the Triangle Group are the Rockley Volcanics, consisting of mafic and ultramafic flows, cumulates and volcaniclastic units. In the Burraga district, these rocks have been metamorphosed to tremolite-chlorite and quartz-feldspar-biotite-amphibole schists (Sheppard *et al.* 1995). Relict clinopyroxene and olivine phenocrysts compositions suggest that these schists were shoshonitic ultramafic cumulates and tuffs.

Alteration of Triangle Group quartzite and schist in the Lucky Draw mine area consists of an early metamorphic stage of medium- to coarse-grained gedrite, cordierite, and staurolite that define the metamorphic fabric of the rock, an intermediate stage of garnet-biotite-chlorite that veins and replaces earlier metamorphic minerals, and a late stage consisting of massive green-brown biotite spatially associated with the Burraga Granodiorite and small dikes. Temperatures of the early metamorphic stage are estimated by Sheppard *et al.* (1995) to be about 600°C based upon mineral equilibria. Temperature of the intermediate garnet-chlorite alteration was calculated to be 62°C based upon Fe-Mg exchange between coexisting mineral pairs.

Mineralisation at Lucky Draw is in the form of Au-As-Bi-Te minerals that are strongly associated with the garnet-chlorite stage of alteration. Identified minerals include ilmenite, arsenopyrite, molybdenite, native gold, maldonite (Au_2Bi), native bismuth, bismuthinite, hedleyite ($Bi_{14}Te_6$), joseite-B ($Bi_{4+x}Te_{2-x}S$), tellurobismuthinite (Bi_2Te_3), and emplectite ($CuBi_2S$) (Sheppard *et al.* 1995). The mineralisation and all stages of alteration are very sulphide poor. Sheppard *et al.* (1995) state that pyrrhotite is the most abundant sulphide and estimate its abundance at less than 0.1 per cent of mineralised sections.

The association of mineralisation with the garnet-chlorite stage of metasomatic alteration suggests that introduction and/or remobilisation of ore elements, Au-As-Bi-Te, occurred after the main phase of penetrative deformation and prior to the biotitisation directly associated with crystallisation and fluid exsolution from the Buragga Granodiorite. This is consistent with other studies, which have suggested that lode gold mineralisation occurs post-peak metamorphism and is synchronous with or slightly earlier than plutonism (*e.g.*, Mueller 1997).

The calculated 538±62°C temperature of garnet-chlorite alteration is higher than the melting temperature of many of the ore minerals, suggesting that fluids circulating during garnet-chlorite alteration, perhaps driven by intrusion of the Buragga Granodiorite, leached ore elements from the adjacent mafic-ultramafic Rockley Volcanics and deposited them by reaction with the iron-rich minerals that are so abundant in the Lucky Draw mine area. Similar conclusions about leaching of mafic/ultramafic rocks during high temperature fluid circulation have been reached by other researchers (*e.g.*, Steven 1993; Noertemann 1997).

Tillicum, British Columbia

Tillicum Mountain is located in south central British Columbia along the northern edge of the east trending Nemo Lakes Belt, a five kilometre wide roof pendant within the Cretaceous Nelson Batholith of upper greenschist- to lower amphibolite-grade metavolcanic and metasedimentary rocks, correlated with the Triassic Elise Formation of the Rossland Group (Peterson 1996). Regional metamorphism is sillimanite-grade at 5.0-6.8 kb (500-680 Mpa) and 630-680°C (Parrish 1981). In the Tillicum area, pressure and temperature conditions are thought to be slightly lower, 4.3-6.3 kb (430-630 Mpa) and 523-568°C, respectively and sillimanite does not occur (Ray *et al.* 1985; Peterson 1996). Metasedimentary rocks consist of thinly banded biotite-muscovite phyllite, spotted biotite schist, and graphitic biotite-muscovite phyllite. Metavolcanic rocks consist of shoshonitic and porphyritic mafic flows, tuffs, breccias, and intercalated argillites (Ray & Spence 1986).

In the vicinity of Tillicum Mountain, the Nelson Batholith consists of the Triassic Goatcanyon-Halifax Creek and mela-diorite stocks. The Goatcanyon-Halifax Creek stock is a medium-grained, equigranular quartz monzonite with an ilmenite to magnetite ratio of 5:1 (Peterson 1996). A marginal phase of the Goatcanyon-Halifax Creek stock is an equigranular mela-diorite or hornblendite containing xenoliths of the Goatcanyon-Halifax Creek stock. Based upon the hornblende geobarometer of Hollister *et al.* (1987), Peterson (1996) estimated pressures of emplacement of 6.3 and 4.3 kb, respectively, for the two plutons. Both stocks are undeformed and thus post-date skarn formation and mineralisation associated with the main shearing event. In addition, the Goatcanyon-Halifax Creek stock truncates the trend of mineralisation in underground workings (Peterson 1996).

Two episodes of folding have been documented in the Tillicum area (Ray *et al.* 1985; Peterson 1996). The first episode consists of southwest striking isoclinal folds and the development of a prominent axial planar schistosity (Peterson

1996). In a second episode, this schistosity was then refolded about a southwest plunging synform, accompanied by shearing. Skarn formation and mineralisation can be placed within this structural framework, because calc-silicate minerals overgrow and cut the D_1 metamorphic fabric in the district. In detail, native gold, native bismuth, and bismuthinite occur in fractures in garnet, gold occurs along cleavage planes in pyroxene, and massive sulphide locally replaces calc-silicate skarn. In addition, skarn minerals and mineralised veins are locally sheared and folded by D_2. Thus, skarn formation and mineralisation is post D_1 and roughly synchronous with D_2. On a local scale, Peterson (1996) suggested that deformation was concentrated along the margins of metavolcanic flows and sills that behaved as competent blocks within a weaker metasedimentary matrix and that hydrothermal/metamorphic fluids were focused along these contacts.

All garnets associated with mineralisation are subcalcic. A small calcareous unit in the North Slope area close to the Goatcanyon-Halifax Creek intrusion has grandite garnets, but this occurrence is not mineralised. Metamorphic garnets and skarn garnets related to mineralisation have similar but distinct compositions. Metamorphic garnets occur in banded schists and gneisses as part of the D_1 penetrative deformation. They are more subcalcic than skarn garnets, containing as little as 6 mole per cent grandite in contrast to a maximum of 64 mole per cent grandite in skarn garnets. Moreover, on average they contain twice as much pyrope (up to 22 mole per cent) and half as much spessartine (as low as 15 mole per cent) as do skarn garnets (as little as 1 mole per cent pyrope and as much as 61 mole per cent spessartine). Both pyroxene and amphibole are the typical calcic varieties that occur in "normal" skarn deposits in a non-regional metamorphic environment. Pyroxene is diopsidic (Hd_{4-42}) and amphibole is mostly within the tremolite-actinolite series (Peterson 1996).

Using the garnet-biotite geothermometer of Ferry & Spear (1978), Peterson (1996) calculated a temperature of 523-568°C for garnet skarn formation, assuming a pressure of 6 kb (600 MPa) as determined for the later emplacement of the Goatcanyon-Halifax Creek stock. This temperature range is consistent with the fluid inclusion trapping temperatures (500-550°C) determined from garnet and quartz associated with mineralisation, using a pressure correction based upon the aforementioned pressure of 6 kb (600 MPa). No daughter minerals were observed in any fluid inclusions and fluid inclusions in quartz yielded salinities of 0.7-3.8 eq. wt. per cent NaCl (Peterson 1996).

The other major alteration type in the Tillicum area is biotite hornfels, which affects all the rock types in the district, except the Goatcanyon-Halifax Creek and mela-diorite intrusions. Depending on the protolith being altered, biotite hornfels can be texturally diverse but always contains biotite, quartz, and K-feldspar and usually is unfoliated. More felsic protoliths tend to have more K-feldspar and more mafic protoliths, more biotite. Some of the biotite alteration of metavolcanic rocks, such as the diorite porphyry flows and sills, is relatively coarse grained with biotite up to several mm. Thus, the term biotite hornfels is not strictly appropriate, but has been retained as a field term because biotite is an essential component and the alteration typically is fine grained and granular.

There are six mineralised zones at Tillicum: Heino-Money, East Ridge, Silver Queen, North Slope, Grizzly, and Arnie Flats. Reserves at East Ridge are 1.4 Mt averaging 7.7 g/t Au. Heino-Money is smaller but much higher grade, with 55,000 tons averaging 33.4 g/t Au. Mineralisation is spatially associated with skarn alteration and shear zones. Some of the shearing predates skarn, as the calc-silicate minerals overgrow the penetrative fabric of the earlier metamorphism. In other cases, sheared quartz veins with coarse-grained visible gold have skarn envelopes and some of the calc-silicate minerals are weakly foliated. Thus, it appears that the Tillicum area has been structurally active for a considerable time and that for at least part of that time, skarn-forming hydrothermal fluids were active during shearing. Veins consist of quartz, calcite, pyroxene, amphibole, clinozoisite, garnet, feldspar, titanite, biotite, and muscovite. Sulphide minerals include major pyrrhotite and pyrite. Minor to trace minerals include native gold, marcasite, native bismuth, bismuthinite, hedleyite, and joesite-B (Peterson 1996).

In general, mineralisation in the Tillicum area is not sulphide-rich. One exception is in the Heino-Money zone where pyroxene-amphibole-calcite skarn is replaced by a vein of massive sulphide consisting of pyrrhotite, sphalerite, galena, boulangerite, arsenopyrite, chalcopyrite, and freibergite. Sulphide replacement ranges from 20 to 95 per cent of the rock, averaging 80 per cent. Values of Pb, Zn, and Ag range up to 7.2 per cent, 39.5 per cent, and 100 oz/t, respectively, with gold values of 0.2 to 1.0 oz/t. Massive sulphide replacement zones also occur in the East Ridge zone and although they are geochemically anomalous in Au, Ag, As, and base metals, they are not ore grade (Peterson 1996).

Navachab, Namibia

The open pit mine at Navachab is located 10 km south of Karibib in the southern central zone of the Damara Orogen (Pirajno & Jacob 1991; Moore and Jacob, 1998). Production is 1.8-1.9 t/y Au from a reserve of 9.75 Mt at an average grade of 2.4 g/t Au from pyroxene-rich zones in skarn formed in metasedimentary rocks of the Okawayo Formation of the Swakop Group (Noertemann 1997). Regionally, the Swakop Group includes the Spes Bona Formation, consisting of schists, calc-silicate rocks, and meta-arkoses, the Karibib Formation, consisting of a basal calc-silicate white marble, a dolomitic brown marble, and a hanging wall gray graphitic marble, the Okawayo Formation consisting of calc-silicate marbles, and the Oberwasser Formation, consisting of siliciclastic units (Steven 1993). Within the Okawayo Formation is a distinctive dark rock, which in the mine is called the marker hornfels, but which geochemically is a metamorphosed, late Damaran, camptonitic lamprophyre (Noertemann 1997). This rock served as a fluid barrier and chemical trap for mineralising fluids. Mineralised skarn is immediately adjacent to the meta-lamprophyre, but the lamprophyre itself is barren of gold (Noertemann 1997).

Puhan (1983) documented metamorphic P-T-conditions ranging from 2.6-3.4 kb (260-340 Mpa) and 555°C-645°C in the central Damara Orogen. Noertemann (1997) showed that the area has been affected by a combination of polyphase folding and late-tectonic brittle-ductile shearing. The D_1 deformation produced East-verging recumbent F_1 folds on a scale of several kilometres. Subsequently, this folding was overprinted by a progressive F_2 folding, as indicated by refolding and intrafolial folds. After that, an isoclinal F_3 folding led to the formation of a weakly NW-verging, upright D_3 anticline on a regional scale, which hosts several gold deposits in the south central Damara Orogen, including Navachab.

Navachab represents a reduced distal and Mn-enriched gold skarn formed in banded, predominantly calcite marble with biotite schist and calc-silicate layers. Regional metamorphism of these rocks produced preferred growth of garnet in pelitic layers and of clinopyroxene in carbonate layers. These early metamorphic garnets are intermediate grandite-pyralspite and the metamorphic pyroxenes are salite with only minor johansenite. Metasomatic skarn, veins and overprints these layered metamorphic occurrences and both garnet and pyroxene are enriched in manganese.

This time progression is clearly visible in garnets which have a metamorphic poikiloblastic core and a younger, inclusion-free margin. Associated amphibole is largely tremolite-actinolite with a minor hastingsite component. Amphibole shows a strong decrease in Mg and increase in Fe^{2+}, Mn^{2+}, and Fe^{3+} from marble to skarn.

The occurrence of graphite in the skarn as a product of decarbonatisation implies very low oxygen-fugacities, consistent with the lack of magnetite and haematite (Noertemann 1997).

The ore mineralisation is distinguished into two main parageneses: pyrrhotite, chalcopyrite, arsenopyrite, molybdenite and sphalerite associated with early skarn formation and a younger one with remobilised pyrrhotite, chalcopyrite, pyrite, native bismuth, bismuthinite, maldonite, and native gold, which is associated with skarn and retrograde amphibole alteration. Noertemann (1997) estimated the P-T conditions of ore formation by geobarometry and geothermometry of sphalerite [2-2.5 kb (200-250 Mpa) and 590 °C] and arsenopyrite (575 ± 15°C). As in other regional metamorphic skarn occurrences, these temperatures and pressures are slightly lower than the peak conditions as determined by Puhan (1983).

Lupin, Northwest Territories, Canada

The Lupin deposit is located 400 km NNE of Yellowknife in the Contwoyto Lake-Point Lake area of the Archean Slave Province of the Canadian Shield, and represents the largest (11.8 Mt at an average grade of 10.0 g/t Au, for a total of 117 t Au) of more than 100 gold occurrences in this area (Bullis *et al.* 1994). The Lupin deposit consists of a series of stratabound replacement orebodies developed in iron formation adjacent to cross-cutting quartz veins. The complexly folded iron formation is intercalated with Archean greywackes and turbidites of the Contwoyto Formation. The deposit occurs within the broad metamorphic aureole of a large granodiorite-granite pluton, the Contwoyto batholith, at a distance of about 1.5 km to the south of the contact. In the mine area, the unmineralised iron formation consists of mesobanded quartz and grunerite and is metamorphosed to upper greenschist facies grade (Lhotka 1988). The cordierite isograd, marking the position of amphibolite-facies metamorphism, passes about 400 m to the north of the mine at surface, but intersects the mine workings at a depth of 550 m (Lhotka & Nesbitt 1989). Garnet-biotite pairs from the Lupin mine indicate a temperature of 600°C at 3 kb (300 MPa), slightly higher than the 575°C determined from cordierite mineral assemblages (Lhotka 1988).

The region has been effected by at least three deformation events (King *et al.* 1988; Relf 1989). The first developed prior to the peak of regional metamorphism and consists of tight isoclinal folds in which the S_1 axial planar cleavage is defined by the alignment of biotite and muscovite. The second deformation phase developed during peak metamorphic conditions and F_2 folds are tight to isoclinal with steep plunges and near-vertical axial planes. F_3 folding created crenulations to earlier folds. The Contwoyto batholith is thought to have intruded during D_3. Quartz veins are abundant in the mine and appear to be localised in fold hinges and may be related to the Lupin Fault, which bounds the orebodies to the southwest (Lhotka 1988). Although not ore grade, the quartz veins locally contain native gold, pyrrhotite, arsenopyrite, and scheelite (Lhotka 1988).

The main ore host at Lupin is an iron formation that has been metamorphosed to an assemblage of grunerite-quartz-magnetite and then later retrograded/sulphidised to include hornblende, ilmenite, and pyrrhotite. In sulphide-rich iron formation, particularly near quartz veins, almost all the grunerite is replaced by hornblende, and arsenopyrite, loellingite, and pyrite are present in addition to pyrrhotite. In zones of very intense alteration/sulphidation, calcic garnet, pyroxene, and actinolite are also present (Lhotka 1988). Locally, garnet amphibolite occurs as lenses within or along the margins of iron formation. The garnet amphibolite consists of almandine garnet-grunerite-chlorite and contains hedenber gitic pyroxene near contacts with iron formation (Bullis *et al.* 1994). Retrograde alteration of these rocks, again associated with quartz veins, includes epidote and actinolite in addition to hornblende.

Lupin contains gold mineralisation associated with calc-silicate alteration that is similar to many Phanerozoic gold skarns. In contrast to some of the Phanerozoic gold skarns associated with regional metamorphism where the protoliths typically contain at least some calcium to form calcic garnet and pyroxene, the iron formation host at Lupin is very calcium poor and Lhotka (1988) determined that Ca was introduced metasomatically by hydrothermal fluids.

The source of the calcium and the ultimate source of the hydrothermal fluids are not known. However, intrusive dikes of felsic to intermediate composition, common in many Phanerozoic skarn deposits, are exposed on the lower levels of the mine. The petrology of these dikes and their relationship to the gold skarn orebodies are not known (Lhotka & Nesbitt 1989).

Nevoria, Western Australia

The Nevoria, Marvel Loch, Big Bell, and other gold skarn deposits in Western Australia are located in the mid-central part of the Archean Yilgarn Craton (Mueller 1997). The skarns are hosted by ultramafic to mafic volcanic rocks or banded iron-formation of the Southern Cross Belt, a narrow greenstone terrane of greenschist to amphibolite grade. They are controlled by NW-trending ductile shear zones and occur within the broad, contact-metamorphic aureole of the Ghooli Dome, a composite granodiorite-granite batholith located to the southwest of a major transcurrent shear zone of more than 300 km strike length. Pressure and temperature estimates for peak metamorphism [4 ± 1 kb (400 Mpa) and 570-610°C], suggest that the present erosion surface in the Southern Cross area and the gold deposits it exposes were at a depth of 10 to 15 km during contact metamorphism and subsequent skarn formation (Mueller 1997).

The region has been affected by multiple episodes of deformation. Symmetrical repetition of units on both sides of the Ghooli Dome suggests an early regional fold with steeply dipping (70-80°S) limbs. This fold is then refolded by the second episode of ductile deformation with north-south-striking axial planes and lineations that plunge 70-80°E. Both generations of folds are cut by brittle-ductile faults that strike N10-25°E and dip 55-65°E; fault striations plunge 10-35°S. These faults cut skarn alteration (Mueller 1997).

The Nevoria gold skarn represents a smaller (0.4 Mt underground, averaging 6.3 g/t Au and 3.15 Mt open pit, averaging 2.7 g/t Au) analogue of the Lupin deposit in Canada (Mueller 1997). The Nevoria deposit is hosted by silicate banded iron-formation within amphibolite-facies greenstones overlying the Ghooli Dome batholith. Deep drill holes have intersected massive pegmatite and biotite granite of the Ghooli Dome at a vertical depth of 250 m below the Nevoria open pits, and post-mineralisation pegmatite dikes cut across the orebodies at the higher levels of the mine (Mueller 1988). The orebodies occur in three steeply dipping horizons of grunerite-quartz iron formation, interbedded with fine-grained ultramafic to mafic amphibolites. Skarn is present in both iron formation and in the ultramafic to mafic volcanic rocks, but gold mineralisation is usually of subeconomic grade in the latter (Mueller 1990). Even in the iron formation, the distribution of gold is not homogeneous because high-grade skarns are localised near contacts with metavolcanic rocks. The central portions of the iron formations are less altered and poorly mineralised.

The skarn orebodies at Nevoria occur as stratabound replacements of grunerite-quartz iron formation adjacent to flatly dipping quartz veins. The major prograde alteration minerals are coarse-grained hedenbergite, locally intergrown with minor calcite and rare scheelite, and medium-grained ferro-actinolite or ferro-hornblende. Minor almandine garnet with ferro-hornblende, biotite, ilmenite, and tourmaline occurs in bands parallel to the iron formation mesobanding, and at the margins of cross-cutting quartz veins. Grandite garnet-pyroxene-actinolitic hornblende veins occur in tholeitic and komatiitic amphibolites, but not in iron formation. Retrograde alteration minerals include Fe-rich chlorite, muscovite, and stilpnomelane. The gangue minerals are intimately intergrown with disseminated to massive pyrrhotite, and minor chalcopyrite, loellingite, arsenopyrite, pyrite, and scheelite. Gold occurs in native form, either enclosed in pyrrhotite or in skarn silicates (*e.g.*, hedenbergite; and is often associated with tsumoite (BiTe), Bi-Te sulphides and maldonite (Au_2Bi), which locally has exsolved to native bismuth and native gold (Mueller 1997).

Skarn is also present in the ultramafic to mafic amphibolites intercalated with the iron formation horizons, but is subeconomic in grade. The most prominent alteration features in the more massive amphibolites are zoned garnet-pyroxene replacement veins with highly irregular boundaries (Mueller 1988). The veins consist of a core of grossular garnet and outer margins of diopside. Minor plagioclase, microcline and scheelite are intergrown with both garnet and pyroxene. The peak fluid temperature during the formation of the zoned skarn veins in the ultramafic to mafic volcanic rocks, is constrained by the reaction clinozoisite + quartz + calcite = grossular to values of 550-580°C, assuming a pressure of 4 kb (400 MPa) and low mole fraction CO_2 (0.03-0.05) in the fluid. Retrograde minerals in the veins include the assemblage clinozoisite + calcite + quartz, filling cracks in grossular garnet, and aggregates of muscovite and prehnite replacing feldspar. Disseminated sulphides are rare, and consist mainly of pyrite and chalcopyrite (Mueller 1990).

The difference between skarn hosted in iron formation and in amphibolite goes beyond mineralogy. Mueller (1997) postulated that iron formation and amphibolite have contrasting physical properties and that, during D_1 folding, contacts between these two rock types accommodated most of the differential slip. These contacts were then the locus of preferential alteration and mineralisation.

In conclusion, gold skarns occur worldwide and in a variety of geologic settings. These deposits share many common features such as biotite hornfels, garnet-pyroxene alteration, clastic- and/or volcaniclastic-rich protoliths, and a Au-As-Bi-Te geochemical signature, but also exhibit significant differences, especially among the four major subdivisions: (1) reduced Au skarns, (2) oxidised Au skarns, (3) magnesian Au skarns, and (4) metamorphic Au skarns. Both reduced and oxidised Au skarns are related to shallow Phanerozoic plutons. Most depth estimates for these systems are $\leq$ 5 km, broadly similar to the general environment of porphyry-type deposits.

Plutons associated with reduced gold skarns tend to be ilmenite-bearing mafic diorites and granodiorites, whereas plutons associated with oxidised gold skarns tend to be more silicic and magnetite-bearing (Meinert 1995; Ray *et al.* 1995). In contrast, magnesian and metamorphic Au skarns do not necessarily occur with associated igneous rocks and they range in age from Archean to Phanerozoic. The formation of skarn in these systems appears to be more dependent on particular host rock compositions and relatively high P-T conditions rather than on the petrochemistry of associated plutons.

Yet even with these fundamental differences, most of the deposits still exhibit biotite hornfels, garnet-pyroxene alteration, and a Au-As-Bi-Te geochemical signature. Thus, it seems that fundamentally similar geochemical processes can occur in what appear to be relatively disparate geological environments. A similar conclusion has been reached for other geologic environments, such as geothermal (Bird *et al.* 1984) and VMS (Galley & Ames 1998) deposits, some of which also exhibit skarn alteration.

The biotite±Kfeldspar (potassic) alteration that surrounds most gold skarns is one of the characteristics of this deposit type. In most cases, the biotite±Kfeldspar alteration forms in relatively fine-grained, clastic host rocks, resulting in a hornfels texture. However, in coarser-grained rocks, particularly igneous rocks, the biotite±Kfeldspar alteration can be relatively coarse-grained, as has been described at Hedley in the reduced skarn class (Ettlinger *et al.* 1992; Ray *et al.* 1996b) and Tillicum in the metamorphic skarn class (Peterson 1996). On the scale of a hand specimen, the biotite±Kfeldspar alteration clearly is metasomatic both in the sense of fluid infiltration via fractures and in the sense of mass transfer, *e.g.*, K is being added to the rock.

However, on the larger deposit scale, much of the biotite±Kfeldspar alteration appears to be more a case of in situ redistribution during progressive mineral reactions, as documented at Beal (Wilkie 1996). In most cases, the original clastic sedimentary rocks contain abundant K-rich argillaceous and arkosic material which has been replaced by pyroxene and then garnet in inner or more intense zones of alteration. Since garnet and pyroxene do not contain K, the K originally present in the host rocks is liberated during skarn formation and may be incorporated into the distal biotite±Kfeldspar alteration. In contrast to other elements such as Fe and Au, which are demonstrably added to the alteration assemblage from magmatically derived hydrothermal fluids (*e.g.*, Zimmerman *et al.* 1992), K in biotite±Kfeldspar alteration appears to be, at least in part, internally derived from the host stratigraphy. This conclusion is supported by the fact that skarns developed in relatively pure limestone do not generally develop biotite hornfels or other K-rich alteration.

If biotite±Kfeldspar hornfels is thought of as the distal alteration zone of gold skarns, then the proximal part is relatively coarse-grained garnet and pyroxene. As documented for numerous deposits, this proximal skarn zone is internally zoned from garnet-dominant close to the pluton or fluid pathway to pyroxene-dominant away from the pluton or fluid pathway (Meinert 1997). The relative proportion of garnet and pyroxene is a complex function of protolith composition, activity of components in the hydrothermal fluid, and overall oxidation state as influenced by magmatic sources, wall-rock composition, and mineral reactions (*e.g.*, Newberry 1991). In a general way this corresponds to the fundamental classification of a given skarn as reduced versus oxidised. The oxidation state of a skarn system is reflected in both the relative proportions of garnet and pyroxene and in the composition of individual mineral phases.

The most reduced skarn assemblages lack ferric iron and contain major amounts of ferrous iron in both garnet and pyroxene. Generally, such compositions only occur in skarns, such as Lupin, Navachab, and Nevoria, formed from Fe-rich rocks at great depth in regional metamorphic terrains. Pressures in these deposits have been estimated at 2-6 kb (200-600 MPa), in contrast to $\leq$ 0.5 kb (50 MPa) in the less reduced skarns associated with hypabyssal intrusions. W skarns provide a link between these two groups in that some of the deeper W skarns with particularly reduced graphitic wall rocks have hedenbergitic pyroxene and Fe-rich subcalcic garnets (Newberry 1983).

Although W skarn garnets do not contain nearly as much almandine component as do the garnets in metamorphic Au skarns, the subcalcic almandine component can be taken as an indirect measure of very low oxidation states. The next most reduced mineral assemblage lacks significant ferric iron in garnet, has major amounts of ferrous iron in pyroxene, but lacks ferrous iron in garnet. Au skarns with such hedenbergitic pyroxene and intermediate grandite garnet include Crown Jewel, Elkhorn, Fortitude, Hedley, and Junction Reefs. The least reduced mineral assemblage is characterised by major amounts of andraditic garnet and diopsidic to salitic pyroxene. This assemblage is characteristic of all the skarns in the oxidised Au skarn category.

Although most sulphide mineralisation occurs at lower temperatures after the main stage of garnet and pyroxene formation, the silicate, oxide, and sulphide mineral association of a given skarn rock tends to reflect the overall oxidation and sulphidation state of the hydrothermal system. This can be represented for Au skarn deposits by considering relevant silicate, oxide, and sulphide mineral equilibria. Oxidised Au skarns typically contain andraditic garnet, diopsidic pyroxene, tremolite-actinolite, pyrite, and haematite. Reduced Au skarns typically contain intermediate grandite garnet, hedenbergitic pyroxene, actinolite-ferro-actinolite, pyrrhotite and magnetite.

Metamorphic Au skarns typically contain subcalcic garnet with significant almandine-spessartine, hedenbergitic pyroxene, grunerite-ferro-actinolite, pyrrhotite and graphite.

The metamorphic Au skarn category provides an important link to gold deposits, typically described as orogenic or mesothermal lode Au deposits, that are associated with shear zones in deep crustal environments (*e.g.*, Mueller & Groves 1991; McCuaig & Kerrich 1994). These deposits are characterised by sericite-carbonate-albite wall rock alteration and lack calc-silicate minerals. Even where the protoliths are particularly Fe-rich, such as iron formation, calc-silicate minerals are lacking (e.g., Williams 1997). The metamorphic Au skarns form in the same general geologic environment and under similar P-T conditions. They differ mainly in the composition of the fluid phase and the resulting metasomatism that stabilises calc-silicate minerals. Mueller & Groves (1991) suggested that skarn-forming metasomatism in shear zones is restricted to deeper, higher temperature occurrences. Other workers have ascribed this metasomatic signature to a particular igneous petrogenesis (*e.g.*, Spooner 1993) but the apparent lack of magmatism in some districts like Navachab suggests that co-genetic magmatism is not required in all cases.

The Au-As-Bi-Te geochemical signature of Au skarns also can be related to geochemical variations involving temperature, $f(O_2)$, and $f(S_2)$. For example there are two main As minerals in gold skarns, arsenopyrite, and loellingite. The assemblage loellingite-pyrrhotite is only stable at a combination of relatively high temperature and relatively low sulphidation state. Such conditions are typical of the very reduced metamorphic gold skarns and loellingite-pyrrhotite is reported from all these deposits. In many deposits, such as Navachab (Noertemann 1997), loellingite is described as rimmed or partially replaced by arsenopyrite, particularly where associated with retrograde alteration of associated silicate minerals. This is consistent with isobaric cooling at low $f(O_2)$ and $f(S_2)$ conditions. In the reduced Au skarns associated with Phanerozoic hyabyssal intrusions, loellingite is less common than arsenopyrite in all deposits and is absent from some. Arsenopyrite in these deposits typically occurs with pyrrhotite and generally is reported to be relatively late in the paragenesis, *e.g.*, at relatively low temperature compared to the early garnet and pyroxene. In the oxidised Au skarns, loellingite is absent and arsenopyrite is not abundant, typically occurring with pyrite and as late cross cutting veins, indicative of relatively high sulphidation states and/or low temperatures.

One of the apparent anomalies of Au skarn mineralogy is the ubiquitous occurrence of Bi, Bi-Au, and Bi-Te minerals with very low melting points, in most cases much lower than the formation temperature of associated silicate, oxide, and sulphide minerals. For example, maldonite (Au_2Bi) unmixes into native Au and native Bi at a temperature of 113°C (Barton & Skinner 1979). Most of the Bi, Bi-Au, and Bi-Te minerals in Au skarns probably were intermediate composition solid solutions at the temperature of initial skarn formation and their present mineralogy represents a long history of unmixing and remobilisation. Particularly in metamorphic environments, it might be predicted that these elements will have a long and varied history during protracted thermal, hydrothermal, and structural events.

In contrast to many other types of gold deposits, such as the epithermal class in which transport of gold as bisulphide complexes at temperatures < 300°C and the mechanical/chemical effects of boiling are critical features, Au skarns contain abundant evidence for high-temperature transport of Au in saline to hypersaline fluids. At the high temperatures documented for many skarn deposits (350-650°C), Au could be transported in high concentrations as chloride complexes (Gammons & Williams-Jones 1995, 1997), whereas at temperatures below 300°C, a thiosulphide complex [$Au(HS)_2^-$] is more likely (Hayashi & Ohmoto 1991). One of the objections to chloride transport of Au in even relatively high-temperature mesothermal Au deposits is the lack of base metals in these deposits even though Fe, Cu, Zn, *etc.* are highly soluble in high-temperature chloride fluids (McCuaig & Kerrich 1994). This objection does not hold for most Au skarns as base metals, particularly Fe, are common.

The problem then becomes more one of getting gold out of chloride solutions rather than into solution. For Au skarns associated with plutons, the inevitable cooling of hydrothermal fluids away from the magma and as the magma itself cools, provides a ready mechanism for depositing gold due to the relatively low solubility of Au chloride complexes below 400°C, and especially below 300°C. Meinert (1989) suggested coupled transport for such Au skarns, with chloride transport at high temperatures and bisulphide transport becoming increasingly important below about 350°C. Due to the broadly isothermal conditions during regional metamorphism, this mechanism is less likely for the regional metamorphic Au skarns.

In these systems, the mineralogical evidence for fluid-wallrock reaction suggests that Au deposition may be due to sulphidation of Fe-rich wallrocks such as at Lupin (Bullis *et al.* 1994) or pH changes due to K metasomatism or reaction with carbonate wallrocks. Alternatively, the relatively low oxidation and sulphidation states of most Au skarn deposits provide a clue that reduction of an oxidised hydrothermal fluid may be a fundamental cause for the precipitation of gold in all of these systems.

Depending on whether one is a "lumper" or a "splitter", Au skarn deposits are either fairly similar since they all contain calc-silicate minerals or they are vastly different because they span such a wide variety of host rocks, intrusive relationships, and details of mineral chemistry. What unites them is a common mineralogy and geochemistry rooted in the definition of skarn. But it is not the individual minerals that provide the key to understanding these fascinating deposits, but rather the underlying geological and chemical processes that control mineral stabilities and control whether a system has the potential for turning into a profitable mine or remaining a mineralogical curiosity. Although relatively young Phanerozoic gold skarns directly associated with shallow intrusions are the best known of the systems described in this review, the presence of skarn-type systems in older rocks and associated with the much larger scale of regional metamorphism, suggests a new target of opportunity for exploration and research. Just as in the case when serious exploration of gold skarn systems was spurred by the dramatic rise in the price of gold during the 1980s, the re-examination of old familiar metamorphic terrains with a fresh eye, a fresh "skarn eye", may yield significant new discoveries and yet another new gold rush for the next century.

Tungsten Skarns

Tungsten skarns are found on most continents in association with calc-alkaline plutons in major orogenic belts. Major reviews of tungsten skarns include Newberry and Einaudi (1981), Newberry and Swanson (1986), Kwak (1987), and Newberry (1998). As a group, tungsten skarns are associated with coarse-grained, equigranular batholiths (with pegmatite and aplite dikes) surrounded by large, high-temperature, metamorphic aureoles. These features are collectively indicative of a deep environment. Plutons are typically fresh with only minor myrmekite and plagioclase-pyroxene endoskarn zones near contacts.

The high-temperature metamorphic aureoles common in the tungsten skarn environment contain abundant calc-silicate hornfels, reaction skarns, and skarnoid formed from mixed carbonate-pelite sequences. Such metamorphic calc-silicate minerals reflect the composition and texture of the protolith and can be distinguished from ore-grade metasomatic skarn in the field and in the laboratory.

Newberry and Einaudi (1981) divided tungsten skarns into two groups: reduced and oxidised types, based on host rock composition (carbonaceous versus hematitic), skarn mineralogy (ferrous versus ferric iron), and relative depth (metamorphic temperature and involvement of oxygenated groundwater).

Early skarn assemblages in reduced tungsten skarns are dominated by hedenbergitic pyroxene and lesser grandite garnet with associated disseminated fine-grained, molybdenum-rich scheelite (powellite). Later garnets are subcalcic (Newberry, 1983) with significant amounts (up to 80 mole per cent) of spessartine and almandine. This subcalcic garnet is associated with leaching of early disseminated scheelite and redeposition as coarse-grained, often vein-controlled, low-molybdenum scheelite. It is also associated with the introduction of sulphides, such as pyrrhotite, molybdenite, chalcopyrite, sphalerite, and arsenopyrite, and hydrous minerals such as biotite, hornblende, and epidote.

In oxidised tungsten skarns, andraditic garnet is more abundant than pyroxene, scheelite is molybdenum-poor, and ferric iron phases are more common than ferrous phases. For example at the Springer deposit in Nevada, garnet is abundant and has andraditic rims, pyroxene is diopsidic, epidote is the dominant hydrous mineral, pyrite is more common than pyrrhotite, and subcalcic garnet is rare to absent (Johnson and Keith, 1991). In general, oxidised tungsten skarns tend to be smaller than reduced tungsten skarns, although the highest grades in both systems typically are associated with hydrous minerals and retrograde alteration.

Copper Skarns

Copper skarns are perhaps the worlds most abundant skarn type. They are particularly common in orogenic zones related to subduction, both in oceanic and continental settings. Major reviews of copper skarns include Einaudi et al. (1981) and Einaudi (1982a,b). Most copper skarns are associated with I-type, magnetite series, calc-alkaline, porphyritic plutons, many of which have co-genetic volcanic rocks, stockwork veining, brittle fracturing and brecciation, and intense hydrothermal alteration. These are all features indicative of a relatively shallow environment of formation. Most copper skarns form in close proximity to stock contacts with a relatively oxidised skarn mineralogy dominated by andraditic garnet. Other phases include diopsidic pyroxene, idocrase, wollastonite, actinolite, and epidote. Haematite and magnetite are common in most deposits and the presence of dolomitic wall rocks is coincident with massive magnetite lodes which may be mined on a local scale for iron. As noted by Einaudi et al. (1981), copper skarns commonly are zoned with massive garnetite near the pluton and increasing pyroxene and finally idocrase and/or wollastonite near the marble contact. In addition, garnet may be colour zoned from proximal dark reddish-brown to distal green and yellow varieties.

Sulphide mineralogy and metal ratios may also be systematically zoned relative to the causative pluton. In general, pyrite and chalcopyrite are most abundant near the pluton with increasing chalcopyrite and finally bornite in wollastonite zones near the marble contact. In copper skarns containing monticellite (e.g. Ertsberg, Irian Jaya, Indonesia, Kyle et al., 1991; Maid of Erin, British Columbia, Meinert unpub. data) bornite-chalcocite are the dominant Cu-Fe sulphides rather than pyrite-chalcopyrite (e.g. Big Gossan, Irian Jaya, Meinert *et al.*, 1997). The largest copper skarns are associated with mineralised porphyry copper plutons.

These deposits can exceed 1 billion tons of combined porphyry and skarn ore with more than 5 million tons of copper recoverable from skarn. The mineralised plutons exhibit characteristic potassium silicate and sericitic alteration which can be correlated with prograde garnet-pyroxene and retrograde epidote-actinolite, respectively, in the skarn. Intense retrograde alteration is common in copper skarns and in some porphyry-related deposits may destroy most of the prograde garnet and pyroxene (e.g. Ely, Nevada; James, 1976).

Endoskarn alteration of mineralised plutons is rare. In contrast, barren stocks associated with copper skarns contain abundant epidote-actinolite- chlorite endoskarn and less intense retrograde alteration of skarn. Some copper deposits have coarse-grained actinolite-chalcopyrite-pyrite-magnetite ores but contain only sparse prograde garnet-pyroxene skarn (e.g. Monterrosas and Ral-Condestable deposits, Peru: Ripley and Ohmoto, 1977; Sidder, 1984; Vidal et al., 1990; Record mine, Oregon, Caffrey, 1982; Cerro de Mercado, Mexico, Lyons, 1988). These deposits provide a link between some copper and iron skarns and deposits with volcanogenic and orthomagmatic affinities.

ZINC SKARNS

Most zinc skarns occur in continental settings associated with either subduction or rifting. They are mined for ores of zinc, lead, and silver although zinc is usually dominant. They are also high grade (10-20 per cent Zn+ Pb, 30-300 g/t Ag). Related igneous rocks span a wide range of compositions from diorite through high-silica granite. They also span diverse geological environments from deep-seated batholiths to shallow dike-sill complexes to surface volcanic extrusions. The common thread linking most zinc skarn ores is their occurrence distal to associated igneous rocks. Major reviews of zinc skarn deposits include Einaudi et al. (1981) and Megaw et al. (1988).

Zinc skarns can be subdivided according to several criteria including distance from magmatic source, temperature of formation, relative proportion of skarn and sulphide minerals, and geometric shape of the ore body. None of these criteria are entirely satisfactory because a magmatic source cannot be identified for some deposits, because most skarns develop over a range of temperatures, and because most large skarn deposits contain both skarn-rich ores and skarn-poor ores in a variety of geometric settings including mantos and chimneys (e.g. Megaw, 1998). Megaw et al. (1988) made the important point that many zinc skarn districts "grade outward from intrusion-associated mineralisation to intrusion-free ores, which suggests that those districts lacking known intrusive relationships may not have been traced to their ends". Similarly, most zinc skarn districts grade outward from skarn-rich mineralisation to skarn-poor ores, veins, and massive sulphide bodies which may contain few if any skarn minerals. Incompletely explored districts may only have some of these zones exposed. The presence of skarn minerals, such as garnet and pyroxene within the system, is important because it indicates a restricted geochemical environment which is entirely distinct from ore types, such as Mississippi Valley-type deposits, which also contain Zn-Pb-Ag ores but which absolutely lack skarn minerals.

Besides their Zn-Pb-Ag metal content, zinc skarns can be distinguished from other skarn types by their distinctive manganese- and iron-rich mineralogy, by their occurrence along structural and lithologic contacts, and by the absence of significant metamorphic aureoles centred on the skarn. Almost all skarn minerals in these deposits can be enriched in manganese including garnet, pyroxene, olivine, ilvaite, pyroxenoid, amphibole, chlorite, and serpentine.

In some deposits, the pyroxene:garnet ratio and the manganese content of pyroxene increase systematically along the fluid flow path (e.g. Groundhog, New Mexico, Meinert, 1987). This feature has been used to identify proximal and distal skarns and proximal and distal zones within individual skarn deposits. A typical zonation sequence from proximal to distal is: altered/endoskarned pluton, garnet, pyroxene, pyroxenoid, and sulphide/oxide replacement bodies (sometimes called mantos and chimneys based upon geometry and local custom). The occurrence of zinc skarns in distal portions of major magmatic/hydrothermal systems may make even small deposits potentially useful as exploration guides in poorly exposed districts. Thus, reports of manganese-rich mineral occurrences may provide clues to districts that have not yet received significant exploration activity. Another evidence of distal hydrothermal alteration related to skarn deposits is the occurrence of hairline fractures and stylolites in sedimentary rocks beyond the limit of calc-silicate minerals. This has been documented for many different skarn deposits (e.g. Meinert et al., 1997) but is particularly common in 0Zn skarns.

Molybdenum Skarns

Most molybdenum skarns are associated with leucocratic granites and range from high grade, relatively small deposits (Azegour, Morocco, Permingeat, 1957) to low grade, bulk tonnage deposits (Little Boulder Creek, Idaho, Cavanaugh, 1978). Numerous small occurrences are also found in Precambrian stable cratons associated with pegmatite, aplite, and other leucocratic rocks (Vokes, 1963). Most molybdenum skarns contain a variety of metals including W, Cu, Zn, Pb, Bi, Sn, and U and some are truly polymetallic in that several metals need to be recovered together in order for the deposits to be mined economically. Mo-W-Cu is the most common association and some tungsten skarns and copper skarns contain zones of recoverable molybdenum.

Most molybdenum skarns occur in silty carbonate or calcareous clastic rocks; Cannivan Gulch, Montana (Darling, 1990) is a notable exception in that it occurs in dolomite. Hedenbergitic pyroxene is the most common calc-silicate mineral reported from molybdenum skarns with lesser grandite garnet (with minor pyralspite component), wollastonite, amphibole, and fluorite. This skarn mineralogy indicates a reducing environment with high fluorine activities. These deposits have not received significant study outside of the Soviet Union and there has not been a modern review since the brief summary by Einaudi et al. (1981).

Tin Skarns

Tin skarns are almost exclusively associated with high-silica granites generated by partial melting of continental crust. Tin skarns can be subdivided according to several criteria including proximal versus distal, calcic versus magnesian, skarn-rich versus skarn-poor, oxide-rich versus sulphide-rich, and greisen versus skarn. Unfortunately, few of these categories are mutually exclusive.

Many large tin skarn systems are zoned spatially from

skarn-rich to skarn-poor (or absent). For example, in the Renison Bell area of Tasmania, Australia there is a single large magmatic/hydrothermal system zoned from a proximal calcic tin skarn with minor cassiterite disseminated in a sulphide-poor garnet-pyroxene gangue to a distal magnesian massive sulphide replacement body containing abundant cassiterite and a complete absence of calc-silicate minerals. The distal massive sulphide ore body (Renison Bell) is a major ore deposit and the proximal skarn body (Pine Hill) has not and probably never will be mined.

Einaudi et al. (1981) emphasised that there is a common thread linking the several types of tin skarn deposits and that is the characteristic suite of trace elements (Sn, F, B, Be, Li, W, Mo, and Rb) in the ore and in associated igneous rocks. This suite distinguishes tin skarns from all other skarn types. Kwak (1987) makes a further distinction in that many tin skarn deposits develop a greisen alteration stage which is superimposed upon the intrusion, early skarn, and unaltered carbonate. Greisen alteration is characterised by high fluorine activities and the presence of minerals like fluorite, topaz, tourmaline, muscovite, grunerite, ilmenite, and abundant quartz. In many cases this greisen-stage alteration completely destroys earlier alteration stages. Of particular importance, greisen-style alteration is absent from all other skarn types.

There are several mineralogical features of tin skarns that should be highlighted. From a mining standpoint, the most important is that tin can be incorporated into silicate minerals, such as garnet, sphene, and idocrase, where it is economically unrecoverable. Dobson (1982) reports garnet containing up to 6 per cent Sn in skarn at Lost River, Alaska. Thus, large deposits such as Moina in Tasmania (Kwak and Askins, 1981), can contain substantial amounts of tin that cannot be recovered with present or foreseeable technology. Extensive retrograde or greisen alteration of early tin-bearing skarn minerals can liberate this tin and cause it to precipitate in oxide or sulphide ore. Thus, the skarn destructive stages of alteration are particularly important in tin skarn deposits. As noted by Kwak

(1987), the most attractive ore bodies occur in the distal portions of large skarn districts where massive sulphide or oxide replacements occur without significant loss of tin in calc-silicate minerals like garnet.

Other Skarn Types

There are many other types of skarn which historically have been mined or explored for a variety of metals and industrial minerals. Some of the more interesting include rare metal and rare earth element enriched skarns (e.g. Kato, 1989; Birkett and Sinclair, 1998). REEs tend to be enriched in specific mineral phases such as garnet, idocrase, epidote, and allanite. Vesuvianite and epidote with up to 20 per cent REE have been found in some gold skarns and zinc skarns (Gemmel et al., 1992; Meinert, unpublished data).

Some skarns contain economic concentrations of REEs and uranium (Kwak and Abeysinghe, 1987; Lentz, 1991, 1998). The Mary Kathleen skarn deposit in Queensland, Australia is unusual in that REEs and uranium daughter minerals in fluid inclusions suggest that these elements can be strongly concentrated in high-temperature hydrothermal fluids (Kwak and Abeysinghe, 1987). This suggests that other metasomatic environments should be examined for possible concentrations of REEs and uranium.

The occurrence of platinum group elements is reported in some skarns (e.g. Knopf, 1942; Korobeynikov et al., 1998). These deposits have not been well documented in the literature and most appear to represent metasomatism of ultramafic rocks (e.g. Yu, 1985). It is difficult to evaluate the abundance of PGEs in different skarn types because PGEs have not been routinely analysed until recently. Geochemical considerations suggest that PGEs could be transported under very acidic, oxidised conditions (Wood, 1989). In the skarn environment such conditions might be reached in the greisen alteration stage of tin skarns. This might be a direction for future research and exploration.

ZONATION OF SKARN DEPOSITS

In most skarns there is a general zonation pattern of proximal garnet, distal pyroxene, and idocrase (or a pyroxenoid such as wollastonite, bustamite, or rhodonite) at the contact between skarn and marble. In addition, individual skarn minerals may display systematic colour or compositional variations within the larger zonation pattern. For example, proximal garnet is commonly dark red-brown, becoming lighter brown and finally pale green near the marble front (e.g. Atkinson and Einaudi, 1978). The change in pyroxene colour is less pronounced but typically reflects a progressive increase in iron and/or manganese towards the marble front (e.g. Harris and Einaudi, 1982). For some skarn systems, these zonation patterns can be "stretched out" over a distance of several kilometres and can provide a significant exploration guide (e.g. Meinert, 1987). Details of skarn mineralogy and zonation can be used to construct deposit-specific exploration models as well as more general models useful in developing grass roots exploration programmes or regional syntheses. Reasonably detailed zonation models are available for copper, gold, and zinc skarns (Meinert, 1997). Other models can be constructed from individual deposits which have been well studied such as the Hedley Au skarn (Ettlinger, 1992; Ray et al., 1993) or the Groundhog Zn skarn (Meinert, 1982).

GEOCHEMISTRY OF SKARN DEPOSITS

Skarn formation spans almost the complete range of potential ore-forming environments. Most geochemical studies of skarn deposits have focused on mineral phase equilibria, fluid inclusions, isotopic investigations of fluid sources and pathways, and determination of exploration anomaly and background levels. Experimental phase equilibria studies are essential for understanding individual mineral reactions. Such studies can be extended using thermodynamic data to include variable compositions). Another approach is to use a self-consistent thermodynamic database to model potential skarn-forming solutions (e.g. Flowers and Helgeson, 1983; Johnson and Norton, 1985; Ferry and Baumgartner, 1987). Fractionation of elements between minerals (e.g. Ca:Mg in carbonate, Bowman et al., 1982; Bowman and Essene, 1984) also can be used to estimate conditions of skarn formation. A general review of phase equilibria applicable to skarn systems is presented by Bowman (1998). A more specialised treatment of the vector representation of skarn mineral stabilities is presented by Burt (1998). Recent work has incorporated standard phase equilibria treatment of skarn mineralogy along with fluid dynamics to model the metasomatic evolution of skarn systems (Dipple and Gerdes, 1998).

Fluid inclusion studies of many ore deposit types focus on minerals such as quartz, carbonate, and fluorite which contain numerous fluid inclusions, are relatively transparent, and are stable over a broad T-P-X range. However, this broad T-P-X range can cause problems in interpretation of fluid inclusion data, because these minerals may grow and continue to trap fluids from early high temperature events through late low temperature events (Roedder, 1984). In contrast, high temperature skarn minerals such as forsterite, diopside, etc. are unlikely to trap later low temperature fluids (beyond the host mineral's stability range) without visible evidence of alteration. Thus, fluid inclusions in skarn minerals provide a relatively unambiguous opportunity to measure temperature, pressure, and composition of skarn-forming fluids.

Much of the skarn fluid inclusion literature prior to the mid-1980's has been summarised by Kwak (1986), especially studies of Sn and W skarn deposits. Such studies have been very useful in documenting the high temperatures (>700°C) and high salinities (>50 wt. per cent NaCl equiv. and multiple daughter minerals) which occur in many skarns. All the skarn types summarised in Meinert (1992) have fluid inclusion homogenisation temperatures up to and exceeding 700°C except for copper and zinc skarns, deposits in which most fluid inclusions are in the 300-550°C range. This is consistent with the relatively shallow and distal geologic settings inferred respectively for these two skarn types.

Salinities in most skarn fluid inclusions are high; documented daughter minerals in skarn minerals include NaCl, KCl, $CaCl_2$, $FeCl_2$, $CaCO_3$, CaF_2, C, $NaAlCO_3(OH)_2$, Fe_2O_3, Fe_3O_4, AsFeS, $CuFeS_2$, and ZnS (Table 2). Haynes and Kesler (1988) describe systematic variations in NaCl:KCl:$CaCl_2$ ratios in fluid inclusions from different skarns reflecting differences in the fluid source and the degree of mixing of magmatic, connate, and meteoric fluids. In general, magmatic fluids have KCl>$CaCl_2$ whereas high-$CaCl_2$ fluids appear to have interacted more with sedimentary wall rocks.

Fluid inclusions can provide direct evidence for the content of CO_2 (both liquid and gas), CH_4, N_2, H_2S and other gases in hydrothermal fluids. Studies of gas phases and immiscible liquids in fluid inclusions typically show a dominance of CO_2, a critical variable in skarn mineral stability. Although no comparative studies have been done, it appears that CH_4 is slightly more abundant than CO_2 in reduced systems like tungsten skarns (Fonteilles et al., 1989; Gerstner et al., 1989) whereas CO_2 is more abundant than CH_4 in more oxidised systems like copper and zinc skarns (Megaw et al., 1988).

Studies of fluid inclusions in specific skarn mineral phases are particularly useful in documenting the temporal and spatial evolution of skarn-forming fluids and how those changes correlate with compositional, experimental, and thermodynamic data (e.g. Kwak and Tan, 1981; Meinert, 1987). Fluid inclusions also provide direct evidence for the temperature and salinity shift in most skarn systems between prograde and retrograde skarn events. For example, most garnet and pyroxene fluid inclusions in iron skarns have homogenisation temperatures of 370->700°C and 300-690°C, respectively, with salinities up to 50 wt. per cent NaCl equivalent, whereas retrograde epidote and crosscutting quartz veins have homogenisation temperatures of 245-250°C and 100-250°C, respectively, with salinities of less than 25 wt. per cent NaCl equivalent.

In gold skarns, prograde garnet and pyroxene homogeni sation temperatures are up to 730°C and 695°C, respectively, with salinities up to 33 wt. per cent NaCl equivalent. In contrast, scapolite, epidote, and actinolite from these skarns have homogenisation temperatures of 320-400°C, 255-320°C, and 320-350°C, respectively. In tungsten skarns, prograde garnet and pyroxene homogenisation temperatures are up to 800°C and 600°C, respectively, with salinities up to 52 wt. per cent NaCl equivalent. In contrast, amphibole and quartz from these skarns have homogenisation temperatures of 250-380°C and 290-380°C, respectively with salinities of 12-28 and 2.5-10.5 wt. per cent NaCl equivalent (data summarised in Meinert, 1992).

Isotopic investigations, particularly the stable isotopes of C, O, H, and S, have been critically important in documenting the multiple fluids present in most large skarn systems (Shimazaki, 1988; Bowman, 1998). The pioneering study of Taylor and O'Neill (1977) demonstrated the importance of both magmatic and meteoric waters in the evolution of the Osgood Mountain W skarns. Bowman et al. (1985) demonstrated that in high temperature W skarns, even some of the hydrous minerals such as biotite and amphibole can form at relatively high temperatures from water with a significant magmatic component.

Specifically, garnet, pyroxene, and associated quartz from the skarn deposits summarised in Meinert (1992) all have 180 values in the +4 to +9 range consistent with derivation from magmatic waters. In contrast, 180 values for sedimentary calcite, quartz, and meteoric waters in these deposits are distinctly different. In most cases, there is a continuous mixing line between original sedimentary 180 values and calculated 180 values for magmatic hydrothermal fluids at the temperatures of prograde skarn formation. Similar mixing is indicated by 13C values in calcite, ranging from typical sedimentary 13C values in limestone away from skarn to typical magmatic values in calcite interstitial to prograde garnet and pyroxene (Brown et al., 1985). Hydrous minerals such as biotite, amphibole, and epidote from different skarn deposits also display 180 and D values ranging from magmatic to local sedimentary rocks and meteoric waters (Layne et al., 1991). Again, mixing of multiple fluid sources is indicated.

Sulphur isotopic studies on a variety of sulphide minerals (including pyrite, pyrrhotite, molybdenite, chalcopyrite, sphalerite, bornite, arsenopyrite, and galena) from the skarn deposits indicate a very narrow range of 34 values, consistent with precipitation from magmatic fluids. For some of the more distal zinc skarns, sulphur isotopic studies indicate that the mineralising fluids acquired some of their sulphur from sedimentary rocks (including evaporites) along the fluid flow path (Megaw et al., 1988).

Overall, stable isotopic investigations are consistent with fluid inclusion and mineral equilibria studies which demonstrate that most large skarn deposits form from diverse fluids, including early, high temperature, highly saline brines directly related to crystallising magma systems (e.g. Auwera and Andre, 1988). In many systems, the highest salinity fluids are coincident with peak sulphide deposition. In addition, at least partial mixing with exchanged connate or meteoric fluids is required for most deposits with the latest alteration events forming largely from dilute meteoric waters.

Even though skarn metal contents are quite variable, anomalous concentrations of pathfinder elements in distal skarn zones can be an important exploration guide. Geochemical studies of individual deposits have shown that metal dispersion halos can be zoned from proximal base metal assemblages, through distal precious metal zones, to fringe Pb-Zn-Ag vein concentrations (e.g. Theodore and Blake, 1975). Anomalies of 10s to 100s of ppm for individual metals can extend for more than 1000 metres beyond proximal skarn zones. Comparison of geochemical signatures among different skarn classes suggests that each has a characteristic suite of anomalous elements and that background levels for a particular element in one skarn type may be highly anomalous in other skarns.

For example, Au, Te, Bi, and As values of 1, 10, 100, and 500 ppm, respectively, are not unusual for gold skarns but are rare to absent for other skarn types (e.g. Meinert et al., 1990; Myers and Meinert, 1991).

Some skarns have a strong geophysical response (Chapman and Thompson, 1984; Emerson, 1986). Almost all skarns are significantly denser than the surrounding rock and therefore may form a gravitational anomaly or seismic discontinuity. This is particularly evident in some of the large iron skarns which may contain more than a billion tons of magnetite (specific gravity, 5.18). In addition, both skarns and associated plutons may form magnetic anomalies (Spector, 1972). Relatively oxidised plutons typically contain enough primary magnetite to form a magnetic high whereas reduced plutons typically contain ilmenite rather than magnetite and may form a magnetic low (Ishihara, 1977). Skarns may form a magnetic high due to large concentrations of magnetite (Chapman et al., 1986) or other magnetic minerals such as high temperature pyrrhotite (Wotruba et al., 1988). Since metasomatism of dolomitic rocks tends to form abundant magnetite, in magnesian skarn deposits a strong magnetic signature may be able to distinguish original protolith as well as the presence of skarn (Hallof and Winniski, 1971; Chermeninov, 1988).

Electrical surveys of skarns need to be interpreted carefully. Either disseminated or massive sulphide minerals may give strong IP, EM, or magnetotelluric responses in skarn (Emerson and Welsh, 1988). However, metasomatism of carbonate rock necessarily involves the redistribution of carbon. The presence of carbonaceous matter, especially if in the form of graphite, can strongly affect electrical surveys. Such carbon-induced anomalies may be distant from or unrelated to skarn ore bodies.

A few skarns contain sufficient uranium and thorium to be detectable by airborne or ground radiometric surveys (e.g. Mary Kathleen, Australia, Kwak and Abeysinghe, 1987). Detailed studies of such deposits demonstrate that relatively small skarns can be detected and that different types of skarns can be distinguished (e.g. Lentz, 1991). Although gravity, magnetic, electrical, and radiometric methods have all been applied to skarn deposits, their use has not been widespread. Because of the variability of skarn deposits, it probably is necessary to tailor specific geophysical methods to individual skarn deposits or types.

Petrogenesis and Tectonic Settings of Skarn Deposits

Most major skarn deposits are directly related to igneous activity and broad correlations between igneous composition and skarn type have been described by several workers (Zharikov, 1970; Shimazaki, 1975,1980; Einaudi et al., 1981; Kwak and White, 1982; Meinert, 1983; Newberry and Swanson, 1986; Newberry, 1987; 1990). Averages of large amounts of data for each skarn type can be summarised on a variety of compositional diagrams to show distinctions among skarn classes. Tin and molydenum skarns typically are associated with high silica, strongly differentiated plutons. At the other end of the spectrum, iron skarns usually are associated with low silica, iron-rich, relatively primitive plutons.

Other important characteristics include the oxidation state, size, texture, depth of emplacement, and tectonic setting of individual plutons. For example, tin skarns are almost exclusively associated with reduced, ilmenite-series plutons which can be characterised as S-type or anorogenic. These plutons tend to occur in stable cratons in which partial melting of crustal material may be instigated by incipient rifting. Many gold skarns are also associated with reduced, ilmenite-series plutons. However, gold skarn plutons typically are mafic, low-silica bodies which could not have formed by melting of sedimentary crustal material. In contrast, plutons associated with copper skarns, particularly porphyry copper deposits, are strongly oxidised, magnetite-bearing, I-type and associated with subduction-related magmatic arcs. These plutons tend to be porphyritic and emplaced at shallow levels in the earth's crust. Tungsten skarns, on the other hand, are associated with relatively large, coarse-grained, equigranular plutons or batholithic complexes indicative of a deeper environment.

Tectonic setting, petrogenesis, and skarn deposits are intimately intertwined. Some modern textbooks use tectonic setting to classify igneous provinces (Wilson, 1989) or different kinds of ore deposits (Sawkins, 1984). This approach has been less successful in describing ore deposits such as skarns which are the result of processes that can occur in almost any tectonic setting. A useful tectonic classification of skarn deposits should group skarn types which commonly occur together and distinguish those which typically occur in specialised tectonic settings. For example, calcic Fe-Cu skarn deposits are virtually the only skarn type found in oceanic island-arc terranes. Many of these skarns are also enriched in Co, Ni, Cr, and Au. In addition, some economic gold skarns appear to have formed in back arc basins associated with oceanic volcanic arcs (Ray et al., 1988). Some of the key features that set these skarns apart from those associated with more evolved magmas and crust are their association with gabbroic and dioritic plutons, abundant endoskarn, widespread sodium metasomatism, and the absence of Sn and Pb. Collectively, these features reflect the primitive, oceanic nature of the crust, wall rocks, and plutons.

The vast majority of skarn deposits are associated with magmatic arcs related to subduction beneath continental crust. Plutons range in composition from diorite to granite although differences among the main base metal skarn types appear to reflect the local geologic environment (depth of formation, structural and fluid pathways) more than fundamental differences of petrogenesis (Nakano et al., 1990). In contrast, gold skarns in this environment are associated with particularly reduced plutons that may represent a restricted petrologic history.

The transition from subduction beneath stable continental crust to post-subduction tectonics is not well understood. Magmatism associated with shallow subduction angles may have more crustal interaction (Takahashi et al., 1980) and floundering of the downgoing slab may result in local rifting. During this stage the magmatic arc may widen or migrate further inland. Plutons are granitic in composition and associated skarns are rich in Mo or W-Mo with lesser Zn, Bi, Cu, and F. Many of these skarns are best described as polymetallic with locally important Au and As.

Some skarns are not associated with subduction-related magmatism. These skarns may be associated with S-type magmatism following a major period of subduction or they may be associated with rifting of previously stable cratons. Plutons are granitic in composition and commonly contain primary muscovite and biotite, dark gray quartz megacrysts, miarolitic cavities, greisen-type alteration, and anomalous radioactivity. Associated skarns are rich in tin or fluorine although a host of other elements are usually present and may be of economic importance. This evolved suite includes W, Be, B, Li, Bi, Zn, Pb, U, F, and REE.

Bibliography

Anthony, John W., Bideaux Richard A., Bladh Kenneth W. and Nichols Monte C.: *Handbook of Mineralogy: Elements, Sulfides, Sulfosalts,* Mineral Data Publishing, 1990.

———— : *Mineralogy: Silica, Silicates,* Mineral Data Publishing, 1995.

———— : *Mineralogy: Halides, Hydroxides, Oxides,* Mineral Data Publishing, 1997.

———— : *Handbook of Mineralogy, Volume IV, Arsenates, Phosphates, Vanadates,* Mineral Data Publishing, 2000.

———— : *Mineralogy of Arizona,* University of Arizona Press, Tucson, 1977.

Artini E.: *Mineral,* Hoepli, 1934.

Aubert, G., Guillemin C. and Pierrot R.: *Précis de Minéralogie,* Masson, 1978.

Bancroft, Peter: *The World's Finest Minerals and Crystals,* Viking Press, New York, 1973.

Bates, Robert Latimer: *Our Modern Stone Age,* W. Kaufmann, Los Altos, 1982.

Bayliss, Peter, Erd Dick C., Mrose Mary E., Sabina Ann P. and Smith Deane K.: *Mineral Powder Diffraction File, Data Book, International Centre for Diffraction Data,* 1986.

Betekhtin, A. G.: *Kurs Mineralogii,* Mir, Moskva, 1968.

Beudant F. S.: *Traité élémentaire de Minéralogie,* Verdière, 1830.

Blackburn, William H. and Dennen William H.: *Encyclopedia of Mineral Names*, Mineralogical Association of Canada, 1997.

Bowen, Oliver E.: *Rocks and Minerals of the San Francisco Bay Region*, University of California Press, Berkeley, 1962.

Chesterman, Charles Wesley: *The Audubon Society Field Guide to North American Rocks and Minerals*, Knopf, New York, 1978.

Clark A. M.: *Hey's Mineral Index*, Natural History Museum Publications, Chapman, 1993.

Dana Edward Salisbury, Ford William E., *A Textbook of Mineralogy*, John Wiley, 1949.

De Michele, Vincenzo: *Minerals, their Beauty and Structure*, Orbis Pub., London, 1976.

Deer, W. A.: *Rock-Forming Minerals*, Wiley, New York, 1962.

Deer, W.A., Howie R.A. and Zussman J.: *An Introduction to the Rock forming Minerals*, Longman, London, 1993.

Desautels, Paul E.: *The Mineral Kingdom*, Madison Square Press, New York, 1968.

———————: *Treasures in the Smithsonian: The Gem Collection*, Smithsonian Institution Press, Washington, 1979.

Eckel, Edwin Butt: *Minerals of Colorado*, Fulcrum Pub., Golden, 1997.

Embrey, Peter G. and Fuller John P.: *A Manual of New Mineral Names 1892-1978*, British Museum, 1983.

Estavillo, William: *Gems and Minerals of California: A Guide to Localities*, Renaissance House, Frederick, 1992.

Fleischer, Michael and Mandarino Joseph A.: *Glossary of Mineral Species 1995*, The Mineralogical Record, 1995.

Gaines, R.W., Skinner H.C.W., Foord E.E., Mason B. and Rosenzweig A.: *Dana's New Mineralogy*, John Wiley, New York, 1995.

Gramaccioli, C.M.: *Minerali alpini e prealpini*, Atlas, 1975.

Grim Ralph E.: *Clay Mineralogy*, McGraw-Hill, London, 1968.

Hausel, W. Dan and Wayne M. Sutherland: *Gemstones and Other Unique Minerals and Rocks of Wyoming: A Field Guide for Collectors*, Wyoming State Geological Survey, Laramie, 2000.

Haüy, René Just: *Traité de Minéralogie*, Bachelier, 1822

Henry, Darold John: *California Gem Trails: A Field Guide for the Collector of California's Gems and Minerals*, L.R. Gordon, Long Beach, 1957.

Hey, Max H.: *A Second Appendix to the Second Edition of An Index of Mineral Species and Varieties Arranged Chemically*, Trustees of the British Museum, 1974.

——————— : *An Index of Mineral Species Arranged Chemically*, British Museum, 1975.

——————— : *An Index of Mineral Species and Varieties arranged chemically*, Trustees of the British Museum, 1963.

Kipfer Alex: *Mineral Index*, Ott Verlag, 1974.

Lacroix A.: *Minéralogie de la France et de ses anciens Territoires d'Outre-mer*, Blanchard, 1964.

Ladoo, Raymond B.: *Nonmetallic Minerals*, McGraw-Hill, New York, 1951.

MacFall, Russell P.: *Minerals and Gems: A Color Treasury for Collectors and Guide to Hunting Locations*, Crowell, New York, 1975.

MacLachlan, Don: *Complete Book of Rocks, Gems and Minerals*, Petersen Pub. Co., Los Angeles, 1968.

Mandarino Joseph A.: *New Minerals 1990-1994*, The Mineralogical Record Inc., 1997.

——————— : *Fleischer's Glossary of Mineral Species 1999*, The Mineralogical Record, 1999.

Merrill, George P. and Foshag William F.: *Minerals from Earth and Sky: The Story of Meteorites, Gems and Gem*, Smithsonian Institution Series, New York, 1934.

Milovsky, A. V. and Kononov O. V.: *Mineralogy,* Mir, 1985.

National Museum of Natural History: *The National Gem Collection,* Smithsonian Institution, Washington, 1997.

Nickel, Ernest H. and Nichols Monte C.: *Mineral Reference Manual,* Van Nostrand Reinhold, 1991.

Oliver E.: *The Care and Conservation of Geological Material: Minerals, Rocks, Meteorites and Lunar Finds,* Oxford University Press, Boston, 1992.

Palache, Berman and Frondel: *Dana's System of Mineralogy,* Wiley & Sons, 1962.

Pearl, Richard Maxwell: *Cleaning and Preserving Minerals,* Earth Science Pub. Co., Colorado Springs, 1982.

Pellant, Chris: *Rocks and Minerals,* Dorling Kindersley, New York, 1992.

Phillips, William R. and Griffen Dana T.: *Optical Mineralogy: The Nonopaque Minerals,* Freeman, 1981.

Picot P. and Johan Z.: *Atlas of Ore Minerals,* BRGM, 1982.

Ramdohr, Paul and Strunz Hugo: *Klockmanns Lehrbuch der Mineralogie,* Enke, 1978.

———————: *The Ore Minerals and their Intergrowths,* Pergamon, 1969.

Ransom, Jay Ellis: *Gems and Minerals of America: A Guide to Rock Collecting,* Harper & Row, New York, 1974.

Reynolds, Robert E.: *Minerals from the Mojave,* San Bernardino County Museum Association, Redlands, 2000.

Robbins, Manuel: *Fluorescence: Gems and Minerals under Ultraviolet Light,* Geoscience Press, Phoenix, 1994.

———————: *Rocks and Minerals,* DK Pub., New York, 1998.

Roberts, Willard L., Campbell Thomas J., Rapp George R. and Wilson Wendell E.: *Encyclopedia of Minerals,* Van Nostrand Reinhold, 1990.

Roberts, Willard L., Rapp George R. and Weber Julius: *Encyclopedia of Minerals,* Van Nostrand Reinhold, 1974.

Rössler Hans J.: *Lehrbuch der Mineralogie,* VEB, 1987.

Sinkankas, John: *Gemstone and Mineral Data Book,* Geoscience Press, Prescott, 1981.

Sofianides, Anna S.: *Gems and Crystals from the American Museum of Natural History,* Simon and Schuster, New York, 1990.

Stalder, H.A., Wagner A., Graeser St. and Stucker P.: *Mineralienlexikon der Schweiz,* Wepf, 1998.

Strunz, Hugo and Nickel Ernest H.: *Strunz Mineralogical Tables,* Schweizerbart'sche, 2001.

Strunz, Hugo: *Mineralogische Tabellen,* Geest & Portig, 1982.

Verbeeck: *Biominerals,* CRC Press, Boca Raton, 1990.

Vozza, Vincent J. and Judith L. Vozza: *Zeolites!: The Misunderstood Minerals,* F. Silva Printing, New Jersey, 1975.

Weiss, Stefan: *Das Grosse Lapis Mineralienverzeichnis,* Weise Verlag, 1994.

William, E.: *Familiar Rocks and Minerals,* Knopf, New York, 1988.

Zim, Herbert Spencer: *Rocks and Minerals: A Guide to Familiar Minerals, Gems, Ores and Rocks,* Golden Press, New York, 1957.

Zoltai, Tibor and Stout James A., *Mineralogy, Concepts and Principles,* MacMillan Publishing Company, 1984.

Index

❑❑❑